Selected Titles in This Series

(See the AMS catalog for earlier titles)

Tensor Products and Independent Sums of $\mathcal{L}_p$-Spaces, $1 < p < \infty$

MEMOIRS

of the
American Mathematical Society

Number 660

Tensor Products and Independent Sums of $\mathcal{L}_p$-Spaces, $1 < p < \infty$

Dale E. Alspach

March 1999 • Volume 138 • Number 660 (third of 4 numbers) • ISSN 0065-9266

American Mathematical Society
Providence, Rhode Island

1991 *Mathematics Subject Classification.*
Primary 46B20, 46E30.

Library of Congress Cataloging-in-Publication Data

Alspach, Dale E. (Dale Edward), 1950–
Tensor products and independent sums of $\mathcal{L}p$-spaces, $1 < p <$ [infinity] /Dale E. Alspach.
p. cm. — (Memoirs of the American Mathematical Society, ISSN 0065-9266 ; no. 660)
On t.p. "[infinity]" appears as the infinity symbol.
"Volume 138, number 660 (third of 4 numbers)."
Includes bibliographical references.
ISBN 0-8218-0961-X (alk. paper)
1. $\mathcal{L}p$ spaces. 2. Tensor products. I. Title. II. Series.
QA3.A57 no. 660
[QA323]
510 s—dc21
[515′.73]
98-53108
CIP

Memoirs of the American Mathematical Society

This journal is devoted entirely to research in pure and applied mathematics.

Subscription information. The 1999 subscription begins with volume 137 and consists of six mailings, each containing one or more numbers. Subscription prices for 1999 are $448 list, $358 institutional member. A late charge of 10% of the subscription price will be imposed on orders received from nonmembers after January 1 of the subscription year. Subscribers outside the United States and India must pay a postage surcharge of $30; subscribers in India must pay a postage surcharge of $43. Expedited delivery to destinations in North America $35; elsewhere $130. Each number may be ordered separately; *please specify number* when ordering an individual number. For prices and titles of recently released numbers, see the New Publications sections of the *Notices of the American Mathematical Society.*

Back number information. For back issues see the *AMS Catalog of Publications.*

Subscriptions and orders should be addressed to the American Mathematical Society, P. O. Box 5904, Boston, MA 02206-5904. *All orders must be accompanied by payment.* Other correspondence should be addressed to Box 6248, Providence, RI 02940-6248.

Memoirs of the American Mathematical Society is published bimonthly (each volume consisting usually of more than one number) by the American Mathematical Society at 201 Charles Street, Providence, RI 02904-2294. Periodicals postage paid at Providence, RI. Postmaster: Send address changes to Memoirs, American Mathematical Society, P. O. Box 6248, Providence, RI 02940-6248.

CONTENTS

ABSTRACT

Two methods of constructing infinitely many isomorphically distinct $\mathcal{L}_p$-spaces have been published. In this article we show that these constructions yield very different spaces and in the process develop methods for dealing with these spaces from the isomorphic viewpoint. We use these methods to give a complete isomorphic classification of the spaces R_p^α, $\alpha < \omega_1$, constructed by Bourgain, Rosenthal, and Schechtman and to show that $X_p \otimes X_p$ is not isomorphic to a complemented subspace of any R_p^α. This latter result is a consequence of a general result concerning complemented embeddings of $X_p \otimes X_p$ into independent sums which shows that the tensor product cannot be broken into a $(p,2)-$sum. As a technical tool we also develop methods for dealing with gliding hump arguments in spaces with bases which have many subsequences which span an isomorph of the space.

1991 *Mathematics Subject Classification.* Primary 46B20 Secondary 46E30.
Key words and phrases. tensor product, projection, complemented, ordinal index.
Research supported in part by NSF grant DMS-9301506
Received by the editor April 17. 1996.

INTRODUCTION

The purpose of this paper is to investigate the relationship among three apparently different constructions of $\mathcal{L}_p$-spaces. We will show that two of the methods produce the same isomorphic classes but that the third method produces a fundamentally different class of spaces. In particular the construction due to Bourgain, Rosenthal and Schechtman [BRS] will be shown to produce different spaces than those Schechtman produced to show that there are infinitely many isomorphically distinct $\mathcal{L}_p$-spaces. In order to explore the gap between the two constructions we resurrect a 1974 construction of $\mathcal{L}_p$-spaces due to the author [A1] that was presented in some seminars at Ohio State but was not published at that time. (See [F] for a complete exposition and related results.) All of the methods of construction make use of Rosenthal's fundamental space X_p and thus have a probabilistic or distributional character that makes the passage to the isomorphic level difficult. One consequence of this work is to show that modifications of the ideas of Rosenthal can be used to work with these more complex spaces within the isomorphic framework.

In Chapter 1 we will review the constructions and the basic properties. First we will describe some of the results in Rosenthal's paper.

THEOREM 0.1. *(Rosenthal's Inequality, [R,Theorem 3] or [JSZ]) Let $2 < p < \infty$. Then there exists a constant K_p depending only on p such that if $f_1, \ldots, f_n$ are independent, mean zero random variables in L_p, then*

$$\frac{1}{2} \max\left\{ \left(\sum_{i=1}^{n} \int \left| f_i \right|^p dx \right)^{1/p}, \left(\sum_{i=1}^{n} \int \left| f_i \right|^2 dx \right)^{1/2} \right\}$$

$$\leq \left(\int \left| \sum_{i=1}^{n} f_i \right|^p dx \right)^{1/p}$$

$$\leq K_p \max\left\{ \left(\sum_{i=1}^{n} \int \left| f_i \right|^p dx \right)^{1/p}, \left(\sum_{i=1}^{n} \int \left| f_i \right|^2 dx \right)^{1/2} \right\}.$$

Using this inequality Rosenthal showed that there was a complemented subspace of L_p which he called X_p that was different (isomorphically) from the other complemented subspaces known at the time. In its sequential form $X_{p,(w_n)}$ is the completion of the space of sequences of real numbers (a_n) with only finitely many a_n non-zero under the norm

$$\|(a_n)\| = \max\left\{ \left(\sum_{i=1}^{\infty} |a_n|^p \right)^{1/p}, \left(\sum_{i=1}^{\infty} |a_n|^2 w_n^2 \right)^{1/2} \right\}$$

where (w_n) is a bounded sequence of positive numbers such that for every $\epsilon > 0$,

$$(*) \qquad\qquad \sum_{w_n < \epsilon} w_n^{2p/(p-2)} = \infty.$$

Using this space and a "bad" l_2-space formed by taking the definition of $X_{p,(w_n)}$ as above except that w_n is a constant independent of n, Rosenthal defined a small list of additional $\mathcal{L}_p$-spaces. The spaces he defined were $B_p = (\sum_k X_{p,(w_n^k)})_p$, where $w_n^k = w_k$ for all n and $\lim w_k = 0$, $B_p \oplus X_p$, $X_p \oplus (\sum l_2)_p$, $X_p \oplus B_p$ and $(\sum X_p)_p$. These spaces, L_p, l_p, $l_2 \oplus l_p$, $(\sum l_2)_p$, and X_p were the only $\mathcal{L}_p$-spaces known at the time of Rosenthal's paper. Some of the isomorphic relations between these spaces were determined by Rosenthal. Others were known to the author but never published. The current state of knowledge of these smaller $\mathcal{L}_p$-spaces can be found in the dissertation of G. Force [F].

One of the most interesting parts of Rosenthal's paper is his proof that the space $X_{p,(w_n)}$ does not depend on the sequence (w_n) as long as $(*)$ is satisfied. We will make use of his ideas in this paper, so we record the basic formulas here.

PROPOSITION 0.2. [R,Lemma 7] *Let* $E_1, E_2, \ldots$ *be a sequence of disjoint finite subsets of* $\mathbb{N}$. *For each* $j \in \mathbb{N}$ *let*

$$f_j = \left(\sum_{n \in E_j} w_n^{2p/(p-2)} \right)^{-1/p} \sum_{n \in E_j} w_n^{2/(p-2)} e_n$$

where (e_n) *is the standard coordinate basis of* $X_{p,(w_n)}$ *and let*

$$f_j^*\left(\sum a_n e_n \right) = \left(\sum_{n \in E_j} w_n^{2p/(p-2)} \right)^{-(p-1)/p} \sum_{n \in E_j} a_n w_n^{2(p-1)/(p-2)}.$$

Then (f_j) *is 1-equivalent to the standard basis of* $X_{p,(w_j')}$ *where*

$$w_j' = \left(\sum_{n \in E_j} w_n^{2p/(p-2)} \right)^{(p-2)/(2p)}$$

for each j *and* $Px = \sum_j f_j^*(x) f_j$ *is a contractive projection onto the closed linear span of* (f_j).

Using Proposition 0.2 Rosenthal showed that every space $X_{p,(w_n)}$ is isomorphic to a complemented subspace of $X_{p,(w_{n,k})}$, where $w_{n,k} \downarrow 0$ as $n \to \infty$ and $w_{n,k}$ does not depend on k, and that $X_{p,(w_{n,k})}$ is complemented in every $X_{p,(w_n)}$ such that (w_n) satisfies $(*)$. Rosenthal then used a special sum of spaces and a version of the Pelczynski decomposition method to show that all of the spaces $X_{p,(w_n)}$ such that (w_n) satisfies $(*)$ are isomorphic.

In Chapter 2 we generalize Rosenthal's methods so that we may consider spaces formed by replacing the scalars in the definition of X_p by a sequence of subspaces of L_p. We prove a result analogous to Proposition 0.2 and develop the tools to use a decomposition method argument similar to Rosenthal's. A crucial notion here is the use of a restricted class of operators which are bounded in the p and 2 norms. This idea has been used previously in working with X_p. (See [A2], [AC] and [JO].) These tools allow us to show that for $\omega \leq \alpha < \omega_1$ the spaces $R_p^{\alpha+k}$, $k \in \mathbb{N}$, are isomorphic.

In order to distinguish the spaces R_p^α for α a limit ordinal, we use some ideas of Schechtman [S] but with a different basic space. Because $(R_p^\alpha)^*$ is isomorphic to R_q^α, where $p^{-1} + q^{-1} = 1$, we can work with the range $1 < p < 2$ or the range $2 < p < \infty$. If $p > 2$, isomorphic embedding is not a sufficiently strong criterion to differentiate the spaces. Indeed, for $p > 2$, $(\sum \ell_2)_p$ is both isomorphic to a subspace of R_p^α and contains a subspace isomorphic to R_p^α, $\omega < \alpha < \omega^2$. Yet, as we will see, there are infinitely many distinct isomorphic types in this collection. On the other hand the situation improves if we consider the range $1 < p < 2$. Here we can proceed in an analogous way to Schechtman [S], however we can not use direct iteration because the argument must work transfinitely and we do not have sufficient control of the constants. Whereas Schechtman used spaces of the form $l_{r_1} \otimes l_{r_2} \otimes \cdots \otimes l_{r_k}$, we use the space D_p originally defined in [A1]. Chapter 3 is devoted to completing the following classification of the spaces R_p^α.

THEOREM. $\{R_p^\alpha : \omega \leq \alpha < \omega_1, \alpha \text{ a limit ordinal}\}$ *forms a complete set of isomorphic types. Moreover, if $\alpha < \beta$ are limit ordinals and $1 < p < 2$, then R_p^β does not embed in R_p^α.*

The remaining chapters of the paper are devoted to the relationship between $X_p \otimes X_p$ and $(p, 2)-$sums of subspaces of L_p. The goal is to show that the two structures are essentially incompatible and thus deduce in Chapter 8 that $X_p \otimes X_p$ is not isomorphic to a complemented subspace of any R_p^α. More precisely, we show

THEOREM 8.3. *Suppose that there is a constant D such that for all n, Y_n is a subspace of L_p with a D-unconditional basis . If T is an isomorphism from $X_p \otimes X_p$ into $(\sum Y_i)_{p,2,(w_n)}$ and P is a projection onto the range of T, then there exists an integer N and a subspace Z of $X_p \otimes X_p$, isomorphic to $X_p \otimes X_p$ such $P_N T|_Z$ is an isomorphism and $P_N T(Z)$ is complemented.*

At this point we will also know that for each $\alpha < \omega_1$, R_p^α is isomorphic to an $(p, 2)$ sum of spaces $R_p^{\alpha_n}$ where for each n, $R_p^{\alpha_n}$ is not isomorphic to R_p^α. If $X_p \otimes X_p$ were isomorphic to a complemented subspace of some R_p^γ, $\gamma < \omega_1$, let α be the smallest such ordinal. Since R_p^α is isomorphic to $(\sum R_p^{\alpha_n})_{p,2}$, Theorem 8.3 implies that $X_p \otimes X_p$ is isomorphic to a complemented subspace of $(\sum_{n=1}^N R_p^{\alpha_n})_{p,2}$, for some $N \in \mathbb{N}$. However this space is isomorphic to $R_p^{\alpha'}$ for some α' such that $\alpha' + \omega \leq \alpha$. This contradicts the choice of α. Thus we get the following result.

COROLLARY 8.4. *For all $\alpha < \omega_1$, $X_p \otimes X_p$ is not isomorphic to a complemented subspace of R_p^α.*

A great deal of technical work is required to reach this point. In Chapter 4 we examine isomorphic embeddings of $X_p \otimes X_p$ into $(p, 2)-$sums of subspaces of L_p and show that there are some restrictions on their behavior. Chapter 5 is devoted to developing methods of passing to subsequences of the natural basis of $X_p \otimes X_p$ which are bases for isomorphs of $X_p \otimes X_p$ so that gliding hump style arguments can be used. The approach used is quite general and may be of independent interest. We prove a general sufficient condition for an operator on $X_p \otimes X_p$ to be an isomorphism on a copy of $X_p \otimes X_p$ in Chapter 6. We also look at the isomorphic types of certain natural subspaces of $X_p \otimes X_p$ such as the span of the diagonal and lower triangle basis vectors. With these methods in hand we start looking at complemented embeddings of $X_p \otimes X_p$ into $(p, 2)-$ sums of subspaces of L_p in Chapter 7. The main technical results are proved there.

In Chapter 9 we make some remarks about directions for further work and list some open problems.

For the most part we use standard notation from Banach space theory as may be found in the books of Lindenstrauss and Tzafriri, [LT], [LTI], [LTII]. We use the expression $a \sim b$ to denote equivalence of numerical quantities up to multiplicative constants, i. e., there exist positive numbers K_1, K_2 such that $K_1^{-1} a \leq b \leq K_2 a$. We will assume that the scalar field is the real numbers throughout. Unless otherwise noted $p > 2$ and $q < 2$ is the conjugate index to p. If F is a set, $|F|$ is its cardinality.

Acknowledgement. We would like to thank the Mathematical Sciences Research Institute at Berkeley for its support during which a portion of this work was completed.

CHAPTER 1

THE CONSTRUCTIONS OF $\mathcal{L}_p$-SPACES

In [S] Schechtman used a simple tensor product to construct infinitely many isomorphically distinct $\mathcal{L}_p$-spaces. This tensor is defined only for subspaces of L_p.

DEFINITION 1.1. Let X and Y be subspaces of $L_p[0,1]$ and define $X \otimes Y$ to be the closed linear span of $\{x(t)y(s) : x \in X \text{ and } y \in Y\}$ in $L_p([0,1] \times [0,1])$ with the usual product measure.

This tensor product depends on the representation of X and Y, a priori. Note that if T and S are bounded operators on $L_p[0,1]$ then we may define $T \otimes S$ on $L_p([0,1] \times [0,1])$ by $[T \otimes S](x \otimes y) = (Tx) \otimes (Sy)$ for all $x,y \in L_p[0,1]$. A straight-forward calculation using the Fubini Theorem shows that $T \otimes S$ is well-defined and that $\|T \otimes S\| \leq \|T\| \cdot \|S\|$. It also follows from standard techniques (integration against the Rademacher functions) that if (x_i) and (y_i) are unconditional basic sequences in $L_p[0,1]$ then $(x_i \otimes y_j)_{i,j}$ is an unconditional basis for $[x_i] \otimes [y_j]$. (See [S] or the proof of Lemma 1.2 below.)

Schechtman defines spaces $\otimes^n X_p = X_p \otimes X_p \otimes \cdots \otimes X_p$ (n-times). The remarks above show that $\otimes^n X_p$ is complemented in $L_p([0,1]^n)$ where the projection is $P \otimes P \otimes \cdots \otimes P$ (n-times) and P is a projection from $L_p[0,1]$ onto X_p. Thus it is immediate that $\otimes^n X_p$ is a $\mathcal{L}_p$-space. Unfortunately in this representation the norm of the projection goes to ∞ with n. Indeed, it was communicated to me by Schechtman from Pisier, that the norm of the projection must be at least as large as the product of the smallest norms of projections onto the factors. To see this we have by [TJ,Lemma 32.3] that for $X \subset Y$ the relative projection constant

$$\lambda(\otimes^n X, \otimes^n Y)) = \sup\{|\mathrm{tr}(u : \otimes^n X \to \otimes^n Y)| : u \in B(\otimes^n Y, \otimes^n Y),$$
$$\nu(u) \leq 1 \text{ and } u(\otimes^n X) \subset \otimes^n X\},$$

where tr denotes the trace and ν is the nuclear norm. Because $\mathrm{tr}(\otimes_1^n P) = \prod_1^n \mathrm{tr}(P)$ and $\nu(\otimes_1^n P) \leq \prod_1^n \nu(P)$, it follows that the relative projection constant of $\otimes^n X_p$ in $L_p([0,1]^n)$ is no better than the nth power of the relative projection constant for X_p in $L_p[0,1]$.

Because of Rosenthal's Inequality it is possible to represent $\otimes^n X_p$ as a sequence space relative to the unconditional basis $(x_{k_1} \otimes x_{k_2} \otimes \cdots \otimes x_{k_n})_{k_j \in \mathbb{N}, 1 \leq j \leq n}$ and explicitly compute a formula for an equivalent sequence space norm. (In [S] the case $n = 2$ is given.) The next lemma will allow us to do the computation inductively.

LEMMA 1.2. *Let* (x_n) *be a normalized sequence of mean zero independent random variables in* $L_p[0,1]$, $2 < p < \infty$, *and let* (y_k) *be an unconditional basic sequence in* $L_p[0,1]$ *and* $Y = [y_k]$. *Then for all* $(a_{n,k})$ *in* $\mathbb{R}$,

$$\left\| \sum_n \sum_k a_{n,k} x_n \otimes y_k \right\|$$

$$\sim \max \left\{ \left(\sum_n \left\| \sum_k a_{n,k} y_k \right\|_p^p \right)^{\frac{1}{p}}, \left(\int \left\| \sum_k [\sum_n a_{n,k} \|x_n\|_2 r_n(\omega)] y_k \right\|_Y^p d\omega \right)^{\frac{1}{p}} \right\},$$

where r_n *is the* n*th Rademacher function.*

Proof. Let $f_n = \sum_k a_{n,k} y_k$ for all n. Then for each t, $(x_n \cdot f_n(t))$ is a sequence of mean zero independent random variables in $L_p[0,1]$ and thus by Rosenthal's Inequality,

$$\frac{1}{2} \max \left\{ \left(\sum_n \int \left| x_n(s) f_n(t) \right|^p ds \right)^{\frac{1}{p}}, \left(\sum_n \int \left| x_n(s) f_n(t) \right|^2 ds \right)^{\frac{1}{2}} \right\}$$

$$\leq \left(\int \left| \sum_n x_n(s) f_n(t) \right|^p ds \right)^{\frac{1}{p}}$$

$$\leq K_p \max \left\{ \left(\sum_n \int \left| x_n(s) f_n(t) \right|^p ds \right)^{\frac{1}{p}}, \left(\sum_n \int \left| x_n(s) f_n(t) \right|^2 ds \right)^{\frac{1}{2}} \right\}.$$

Therefore

$$\left(\iint \left| \sum_n x_n(s) f_n(t) \right|^p ds\, dt \right)^{\frac{1}{p}}$$

$$\sim \left(\max \left\{ \sum_n \iint \left| x_n(s) f_n(t) \right|^p ds\, dt, \int \left(\sum_n \int \left| x_n(s) f_n(t) \right|^2 ds \right)^{\frac{p}{2}} dt \right\} \right)^{\frac{1}{p}}$$

$$\sim \max \left\{ \left(\sum_n \|x_n\|_p^p \|f_n\|_p^p \right)^{\frac{1}{p}}, \left(\int \left(\sum_n \|x_n\|_2^2 |f_n(t)|^2 \right)^{\frac{p}{2}} dt \right)^{\frac{1}{p}} \right\}$$

$$\sim \max \left\{ \left(\sum_n \left\| \sum_k a_{n,k} y_k \right\|_p^p \right)^{\frac{1}{p}}, \left(\int \left(\sum_n \|x_n\|_2^2 \left| \sum_k a_{n,k} y_k(t) \right|^2 \right)^{\frac{p}{2}} dt \right)^{\frac{1}{p}} \right\}$$

$$\sim \max \left\{ \left(\sum_n \left\| \sum_k a_{n,k} y_k \right\|_p^p \right)^{\frac{1}{p}}, \left(\iint \left| \sum_n \|x_n\|_2 \sum_k a_{n,k} y_k(t) r_n(\omega) \right|^p d\omega\, dt \right)^{\frac{1}{p}} \right\}$$

(by Khintchin's Inequality)

$$\sim \max \left\{ \left(\sum_n \left\| \sum_k a_{n,k} y_k \right\|_p^p \right)^{\frac{1}{p}}, \left(\int \left\| \sum_k \left[\sum_n a_{n,k} \|x_n\|_2 r_n(\omega) \right] y_k \right\|_Y^p d\omega \right)^{\frac{1}{p}} \right\}.$$

$\square$

Using this lemma we can now compute the sequence space norm of the tensor product of finitely many copies of X_p. Below we use the convention that $\prod_{s \in \emptyset} f(s) = 1$.

PROPOSITION 1.3. *For each i, $1 \leq i \leq n$ let $(x_k^i)_{k \in \mathbb{N}}$ be a sequence of normalized mean zero independent random variables in $L_p[0,1]$ and let $w_k^i = \|x_k^i\|_2$ for all i and k. Then for all $(a(k_1, \ldots, k_n))_{(k_1, \ldots, k_n) \in \mathbb{N}^n}$ in $\mathbb{R}$,*

$$\left\| \sum_{k_1} \cdots \sum_{k_n} a(k_1, \ldots, k_n) x_{k_1}^1 \otimes \cdots \otimes x_{k_n}^n \right\| \sim \max_F \left\{ \left(\sum_{(k_s)_{s \in F}} \left(\sum_{(k_s)_{s \notin F}} a(k_1, \ldots, k_n)^2 W^2(F, k_1, \ldots, k_n) \right)^{\frac{p}{2}} \right)^{\frac{1}{p}} \right\},$$

where the maximum is taken over all subsets F of the first n natural numbers, and $W^2(F, k_1, \ldots, k_n) = \prod_{s \notin F} (w_{k_s}^s)^2$.

Proof. The proof is by induction on the number of factors, n. For $n = 1$ the assertion is immediate from Rosenthal's Inequality. Now assume it for n factors and consider $n + 1$ factors. By Lemma 1.2 with (y_k) replaced by $(x_{k_1}^1 \otimes \cdots \otimes x_{k_n}^n)_{(k_1, \ldots, k_n) \in \mathbb{N}^n}$ and the inductive hypothesis, we have that

$$\left\| \sum_{k_{n+1}} \sum_{k_1} \cdots \sum_{k_n} a(k_1, \ldots, k_n, k_{n+1}) x_{k_1}^1 \otimes \cdots \otimes x_{k_n}^n \otimes x_{k_{n+1}}^{n+1} \right\|$$

$$\sim \max \left\{ \left(\sum_{k_{n+1}} \max_{F \subset \{1,2,\ldots,n\}} \left\{ \left(\sum_{(k_s)_{s \in F}} \left(\sum_{(k_s)_{s \notin F}} a(k_1, \ldots, k_n, k_{n+1})^2 W^2(F, k_1, \ldots, k_n) \right)^{\frac{p}{2}} \right)^{\frac{1}{p}} \right\}^{p} \right)^{\frac{1}{p}}, \right.$$

$$\left. \left(\int \left\| \sum_{k_1, \ldots, k_n} \left[\sum_{k_{n+1}} a(k_1, \ldots, k_n, k_{n+1}) \|x_{k_{n+1}}^{n+1}\|_2 r_{k_{n+1}}(\omega) \right] x_{k_1}^1 \otimes \cdots \otimes x_{k_n}^n \right\|_p^p d\omega \right)^{\frac{1}{p}} \right\}.$$

Interchanging the summation over k_{n+1} and the $\max$ in the first expression produces the required expressions for which a subset of $\{1, \ldots, n+1\}$ would contain $n+1$. Next we will use the following consequence of Kahane's Inequality [W,III.A.18] and the inductive hypothesis to rewrite the second expression and obtain the others.

$$\int \left| \sum_n \left[\sum_k a_{n,k} r_k(\omega) \right]^2 \right|^{p/2} d\omega = \int \left\| \sum_k \left[\sum_n a_{n,k} r_k(\omega) e_n \right] \right\|_{\ell_2}^p d\omega$$

$$\sim \left(\int \left\| \sum_k \left[\sum_n a_{n,k} r_k(\omega) e_n \right] \right\|_{\ell_2}^2 d\omega \right)^{\frac{p}{2}}$$

$$= \left(\sum_n \sum_k a_{n,k}^2 \right)^{\frac{p}{2}},$$

where (e_n) is the usual unit vector basis of ℓ_2.

$$\left(\int \left\| \sum_{k_1,\dots,k_n} \left[\sum_{k_{n+1}} a(k_1,\dots,k_n,k_{n+1}) \| x^{n+1}_{k_{n+1}} \|_2 r_{k_{n+1}}(\omega) \right] x^1_{k_1} \otimes \cdots \otimes x^n_{k_n} \right\|^p_p \, d\omega \right)^{\frac{1}{p}}$$

$$\sim \left(\int \left(\max_F \left\{ \left(\sum_{(k_s)_{s \in F}} \left(\sum_{(k_s)_{s \notin F}} \left[\sum_{k_{n+1}} a(k_1,\dots,k_n,k_{n+1}) \| x^{n+1}_{k_{n+1}} \|_2 r_{k_{n+1}}(\omega) \right]^2 \right. \right. \right. \right. \right.$$

$$\left. \left. \left. \left. W^2(F,k_1,\dots,k_n) \right)^{\frac{p}{2}} \right)^{\frac{1}{p}} \right\} \right)^p \, d\omega \right)^{\frac{1}{p}}$$

$$\sim \max_F \left\{ \left(\sum_{(k_s)_{s \in F}} \int \left(\sum_{(k_s)_{s \notin F}} \left[\sum_{k_{n+1}} a(k_1,\dots,k_n,k_{n+1}) \| x^{n+1}_{k_{n+1}} \|_2 r_{k_{n+1}}(\omega) \right]^2 \right. \right. \right.$$

$$\left. \left. \left. W^2(F,k_1,\dots,k_n) \right)^{\frac{p}{2}} \, d\omega \right)^{\frac{1}{p}} \right\}$$

(by interchanging the *max* and the integral)

$$\sim \max_F \left\{ \left(\sum_{(k_s)_{s \in F}} \left(\sum_{(k_s)_{s \notin F}} \right. \right. \right.$$

$$\left. \left. \left. \sum_{k_{n+1}} a(k_1,\dots,k_n,k_{n+1})^2 \| x^{n+1}_{k_{n+1}} \|^2_2 W^2(F,k_1,\dots,k_n) \right)^{\frac{p}{2}} \right)^{\frac{1}{p}} \right\}.$$

Because $\| x^{n+1}_{k_{n+1}} \|^2_2 W^2(F,k_1,k_2,\dots,k_n) = W^2(F,k_1,\dots,k_n,k_{n+1})$, the proof is complete. $\square$

Next we will briefly describe the construction of $\mathcal{L}_p$-spaces given in [BRS]. Our exposition is slightly different, but the basic ideas are the same.

DEFINITION 1.4. Suppose that for each $n \in \mathbb{N}$, X_n is a subspace of $L_p(\Omega_n, \mu_n)$ for some probability measure μ_n. Let $\Omega = \prod_{n=1}^\infty \Omega_n$ with the product measure $\mu = \prod \mu_n$, and for each n let p_n be canonical map from Ω onto Ω_n and $j_n(f) = f \circ p_n$ for all $f \in \mathcal{L}_p(\Omega_n, \mu_n)$. Let X_0 be the space of constant functions on Ω and j_0 be the inclusion of X_0 into $L_p(\Omega, \mu)$. Let $(\sum X_n)_I$ denote the closed linear span of $\cup_{n=0}^\infty j_n(X_n)$ and $(\sum' X_n)_I$ denote the closed linear span of $\cup_{n=1}^\infty j_n(X_n)$ in $L_p(\Omega, \mu)$. We will call $(\sum' X_n)_I$ the *independent sum* of $\{X_n\}_{n=1}^\infty$ and $(\sum X_n)_I$ the *complete independent sum* of $\{X_n\}_{n=1}^\infty$.

Remark 1.5. We will always choose the spaces $(X_n)_{n \geq 1}$ to be contained in the mean zero functions. This will guarantee (See Lemma H.) that the complete independent sum has a natural unconditional decomposition into the spaces $(X_n)_{n \geq 0}$. In [BRS] the independent sum was used with spaces, (X_n), containing the constants and thus the independent summands were not necessarily direct summands.

In the construction of $\mathcal{L}_p$-spaces in [BRS] a finite ℓ_p sum is used. In this exposition we replace that approach by using the tensor product. For that purpose it is convenient to introduce the notation L^n_p for the space $L_p([0,1], \mathcal{D}_n, \lambda)$

where $\mathcal{D}_n$ is the σ-algebra generated by the intervals $I_k^n = [k2^{-n}, (k+1)2^{-n})$ for $k = 0, 1, \ldots, 2^n - 1$ and λ is Lebesgue measure. (Thus L_p^n is isometric to $\ell_p^{2^n}$.)

For each $\alpha < \omega_1$ we will define a subspace R_p^α of $L_p(\mu)$ for some probability measure μ. The procedure is inductive. Let $R_p^0 = L_p^0$, the space of constant functions. Now suppose that R_p^α has been defined. Define $R_p^{\alpha+1} = L_p^1 \otimes R_p^\alpha$. For a limit ordinal β let $R_p^\beta = (\sum_{n=1}^\infty R_{p,0}^{\alpha_n})_I$ where $R_{p,0}^{\alpha_n}$ is the set of mean zero functions in $R_p^{\alpha_n}$ and $(\alpha_n)_{n \in \mathbb{N}}$ is an enumeration of the ordinals γ, $0 < \gamma < \beta$.

Remark 1.6. Note that because $L_p^n \otimes L_p^m$ is isometric to L_p^{m+n}, $R_p^{\alpha+k}$ is isometric to $L_p^k \otimes R_p^\alpha$, for any $\alpha < \omega_1$. In the definition of R_p^β for β a limit ordinal a more usual definition would be to let α_n increase to β and let $R_p^\beta = (\sum R_p^{\alpha_n})_I$. However no isomorphic theory of complete independent sums was developed in [BRS], so at this stage we do not know whether or not these spaces are all isomorphic.

CHAPTER 2

ISOMORPHIC PROPERTIES OF
$(p,2)$–SUMS AND THE SPACES R_p^α

In [BRS] it was shown by use of an ordinal index that uncountably many of the spaces R_p^α are isomorphically distinct. Unfortunately the nature of the proof is such that it does not provide any additional information on which $\alpha's$ yield isomorphically different spaces. One consequence of the results in this paper is that we will see which ones are in fact different in a direct fashion.

We will now use Rosenthal's inequality to get a little more information about the spaces R_p^α for α a limit ordinal. The approach used here is similar to that in [A1] and [F].

LEMMA 2.1. *For each $n \in \mathbb{N}$ let X_n be a subspace of $L_p(\Omega_n, \mu_n)$ which contains the constants and let $X_{n,0}$ denote the the subspace of X_n of all mean 0 functions in X_n. Then $(\sum X_n)_I$ is isomorphic to $(\sum' X_{n,0})_I \oplus L_p^0$. Consequently, $(\sum X_n)_I$ is isomorphic to the space*

$$Z = \{(f_n)_{n=0}^\infty : f_n \in X_{n,0} \text{ for } n \in \mathbb{N}, f_0 \in L_p^0,$$
$$\text{and } \|(f_n)\| = \max\{(\sum \|f_n\|_p^p)^{1/p}, (\sum \|f_n\|_2^2)^{1/2}\} < \infty\}.$$

Proof. For each $n \in \mathbb{N}$, $X_n = X_{n,0} \oplus [1_{\Omega_n}]$. By definition

$$f \xmapsto{T} (\textstyle\int f dx) 1_\Omega$$
$$\mathrm{Ker}\, T = X_{n,0}$$

$$(\sum X_n)_I = [\cup_{n=0}^\infty j_n X_n] = [\cup_{n=1}^\infty j_n X_{n,0} \cup \{j_n(1_{\Omega_n}) : n = 0, 1, 2, \dots\}].$$

Because $j_n(1_{\Omega_n}) = 1_\Omega$ for every n, it follows that $(\sum X_n)_I = [\cup j_n(X_{n,0})] \oplus [1_\Omega] = (\sum' X_{n,0})_I \oplus L_p^0$. The final assertion is an immediate consequence of Rosenthal's Inequality. $\square$

We will have frequent use for spaces such as Z above so we will use the following notation for norm that occurs there.

DEFINITION 2.2. Let (X_n) be a sequence of subspaces of $L_p(\Omega, \mu)$ for some probability measure μ, and let (w_n) be a sequence of real numbers, $0 \le w_n \le 1$. For any sequence (x_n) such that $x_n \in X_n$ for all n, let

$$\|(x_n)\|_{p,2,(w_n)} = \max\{(\sum \|x_n\|_p^p)^{1/p}, (\sum \|x_n\|_2^2 w_n^2)^{1/2}\}$$

and let

$$X = (\sum X_n)_{p,2,(w_n)} = \{(x_n) : x_n \in X_n \text{ for all } n \text{ and } \|(x_n)\|_{p,2,(w_n)} < \infty\}.$$

We will say that X is the $(p, 2, (w_n))$-sum of $\{X_n\}$. In the special case that $w_n = 1$ for all n, we will sometimes write $(\sum X_n)_{p,2}$ instead of $(\sum X_n)_{p,2,(1)}$. Also, we define

$$\|(x_n)\|_p = (\sum \|x_n\|_p^p)^{1/p} \text{ and } \|(x_n)\|_2 = (\sum \|x_n\|_2^2 w_n^2)^{1/2}\}.$$

In order to deal with these spaces with a norm defined in terms of an L_2 norm and an L_p norm and with similar subspaces of L_p, we will use the terms $(p, 2)-$ bounded, $(p, 2)-$isomorphism, etc., to indicate that the map is bounded, is an isomorphism, etc., in both norms. For example, if T is a map from $(\sum X_n)_{p,2,(w_n)}$ into L_p, we would say that T is $(p, 2)-$bounded, if there exists a constant K such that $\|T(x_n)\|_2 \leq K(\sum \|x_n\|_2^2 w_n^2)^{1/2}$ and $\|T(x_n)\|_p \leq K\|(x_n)\|_{p,2,(w_n)}$ for all $(x_n) \in (\sum X_n)_{p,2,(w_n)}$. Note that this is slightly different than the usage of this terminology in other papers, [AC], [JO]. This concept of a $(p, 2)-$bounded operator is actually implicit in [R] where it is used in estimating the norms of operators and in proving that the spaces $X_{p,(w_n)}$, where (w_n) satisfies (*) are all isomorphic. Rosenthal actually shows that the spaces are $(p, 2)-$isomorphic.

Remark 2.3. By Rosenthal's Inequality one can produce a subspace of L_p isomorphic to the $(p, 2, (w_n))$-sum of (X_n) in the following way. First symmetrize each space X_n by mapping $f \in X_n$ to $Sf \in L_p(\Omega', \nu)$ where $\Omega' = \Omega \times \{0, 1\}$ and $\nu = \mu \times (\delta_0 + \delta_1)/2$, by $Sf(\omega, j) = f(\omega)(-1)^j$. Next apply an isometry J_n from $L_p(\Omega', \nu)$ into $L_p(\Omega' \times [0, 1], \nu \times \lambda)$, where λ is normalized Lebesgue measure on $[0, 1]$ to adjust the ratio between the L_p and L_2 norms, where $J_n f(\omega', \cdot) = f(\omega')1_{[0,\gamma_n]}\gamma_n^{-1/p}$ where $\gamma_n = w_n^{2p/(p-2)}$. For each $n \in \mathbb{N}$ let $\Omega_n = \Omega' \times [0, 1]$ and $\nu_n = \nu \times \lambda$ and identify the target space of J_n with $L_p(\Omega_n, \nu_n)$. Finally let $\Omega_\infty = \prod_{n=1}^\infty \Omega_n$, $\mu_\infty = \prod_{n=1}^\infty \nu_n$ and π_n denote the natural projection from Ω_∞ onto Ω_n. Then for each $n \in \mathbb{N}$, $T_n f = (J_n Sf) \circ \pi_n$ is an isometry from X_n into $L_p(\Omega_\infty, \mu_\infty)$ such that for any sequence (f_n) with $f_n \in X_n$ for all n, $(T_n f_n)$ is a sequence of independent mean zero random variables. By Rosenthal's inequality $\|\sum T_n f_n\|_p$ is equivalent to $\|(T_n f_n)\|_{p,2,(1)}$ and by the definition of T_n we have that $\|(T_n f_n)\|_{p,2,(1)} = \|(f_n)\|_{p,2,(w_n)}$ because $\|T_n f_n\|_2 = \|f_n\|_2 w_n$ for all $n \in \mathbb{N}$. Thus $[T_n X_n]$ is $(p, 2)-$isomorphic to the $(p, 2, (w_n))$-sum of $\{X_n\}$.

Let us now examine the spaces $R_p^{\omega k}$, for $k = 1, 2, \ldots$. First let $k = 1$. For each $n \in \mathbb{N}$, $L_p^n = [1] \oplus [1_{I_k} - 2^{-n}1_{I_0} : k = 1, 2, \ldots, 2^n - 1]$, where $I_k^n = [k2^{-n}, (k+1)2^{-n})$, for $k = 0, 1, \ldots, 2^n - 1$, and $n \in \mathbb{N}$. Because of Rosenthal's Inequality and Lemma 2.1 it is natural to compare R_p^ω with $(\sum L_p^n)_{p,2,(1)}$. Let $1_n = (x_k) \in (\sum L_p^n)_{p,2,(1)}$, where $x_n = 1_{[0,1]}$ and $x_k = 0$ for $k \neq n$. Observe that the operator $P(x_n) = ((\int x_n)1_n)$ on $(\sum L_p^n)_{p,2,(1)}$ is a (p,2)-norm 1 projection onto $[1_n : n \in \mathbb{N}]$ with kernel $(\sum L_p^n 0)_{p,2,(1)}$. Also we have that $[1_n : n \in \mathbb{N}]$ is $(p, 2)-$isometric to $X_{p,2,(1)}$ which is isomorphic (but not $(p, 2)-$isomorphic) to ℓ_2.

Let us explicitly compute the norm of an element $x \in (\sum L_p^n)_{p,2,(1)}$. Let

$$x = (\sum_{k=0}^{2^n-1} a_k^n 1_{I_k^n})_{n \in \mathbb{N}}.$$

Then

$$\|x\|_{p,2,(1)} = \max\{(\sum_{n=1}^{\infty} \sum_{k=0}^{2^n-1} |a_k^n|^p 2^{-n})^{1/p}, (\sum_{n=1}^{\infty} \sum_{k=0}^{2^n-1} |a_k^n|^2 2^{-n})^{1/2}\}.$$

If we define $b_k^n = a_k^n 2^{-n/p}$ and $w_k^n = 2^{-n(p-2)/(2p)}$ for each n, k, then

$$\|x\| = \max\{(\sum_{n=1}^{\infty} \sum_{k=0}^{2^n-1} |b_k^n|^p)^{1/p}, (\sum_{n=1}^{\infty} \sum_{k=0}^{2^n-1} |b_k^n|^2 (w_k^n)^2)^{1/2}\}.$$

Thus $(\sum L_p^n)_{p,2,(1)}$ is $(p,2)-$isometric to $X_{p,(w_k^n)}$, an isomorph of X_p. By Rosenthal's argument $X_p \oplus_{p,2} X_{p,2,(1)}$ is $(p,2)-$isomorphic to X_p, it follows that $(\sum L_p^n)_I$ is $(p,2)-$isomorphic to $(\sum L_p^n)_{p,2,(1)}$.

Now we will prove that the spaces $R_p^{\omega k}$ are $(p,2)-$isomorphic to iterated $(p,2)-$sums of X_p.

THEOREM 2.4. *For any countable ordinal α and $j \in \mathbb{N}$, $R_p^{\alpha+\omega j}$ is $(p,2)-$ isomorphic to*

$$\underbrace{(\sum(\sum \cdots (\sum R_p^{\alpha})_{p,2,(w_{n,k})}) \cdots)_{p,2,(w_{n,k})})_{p,2,(w_{n,k})}}_{j \text{ times}},$$

$1<2<3<\cdots<\omega<\omega+1<\omega+2<\cdots<\omega 2<\omega 2+1<\omega 2+2<\cdots<\omega 3<\cdots$

where $w_{n,k} = 2^{-n(p-2)/(2p)}$ for $0 \le k < 2^n, n = 1, 2, \ldots$.

Before we begin the proof we need to make note of some properties of $(p,2)-$ isomorphisms. Below, given a sequence of operators (T_n), $T_n : X_n \to Y_n$, we will use the notation $\sum \oplus T_n$ to denote the operator from $(\sum X_n)$ to $(\sum Y_n)$ which is defined by $T((x_n)_{n\in\mathbb{N}}) = (T_n(x_n))_{n\in\mathbb{N}}$, for all finitely non-zero sequences with $x_n \in X_n$ for all n.

LEMMA 2.5. *If $\{X_n\}$ and $\{Y_n\}$ are two sequences of subspaces of $L_p(\Omega, \mu)$ for some probability measure μ, and there is a constant K and $(p,2)-$continuous operators $\{T_n\}$ such that $\|T_n\|_{p,2} \le K$ for all $n \in \mathbb{N}$ then the operator $T = \sum \oplus T_n$ is a K $(p,2)-$bounded operator from $(\sum X_n)_{p,2,(w_n)}$ into $(\sum Y_n)_{p,2,(w_n)}$. Consequently,*

(1) *if each T_n is a $(p,2)-$isomorphism and $\|T_n^{-1}\|_{p,2} \le K$ as well then T is a $(p,2)-$isomorphism.*

(2) *if each T_n is a projection, then $\sum \oplus T_n$ is a $(p,2)-$bounded projection onto*

$$(\sum T_n X_n)_{p,2,(w_n)}.$$

The proof of Lemma 2.5 is straightforward and we leave it to the reader. The next lemma is similar to Proposition 0.2 except that the scalars are replaced by a subspace of L_p.

LEMMA 2.6. *Suppose that $\{X_n\}$ is a sequence of subspaces of $L_p(\Omega, \mu)$ for some probability measure μ, Y is a subspace of $L_p(\Omega, \mu)$, and K is a constant such that for every $n \in \mathbb{N}$ there is a projection P_n from X_n onto a subspace Y_n such that $\|P_n\|_{p,2} \leq K$ and there is a $(p,2)-$isomorphism T_n from Y onto Y_n with $\max\{\|T_n\|_{p,2}, \|T_n^{-1}\|_{p,2}\} \leq K$. Then, if (w_n) is any sequence in $[0,1]$ with $\sum w_n^{2p/(p-2)} < \infty$,*

$$Z = [(w_n^{2/(p-2)} T_n y) : y \in Y] \subset (\sum X_n)_{p,2,(w_n)}$$

is $(p,2)-$isomorphic to Y with the norm

$$\max\{\|y\|_p, \|y\|_2 (\sum w_n^{2p/(p-2)})^{(p-2)/2p}\}$$

and $(p,2)-$complemented in $(\sum X_n)_{p,2,(w_n)}$. Moreover the norms of the operators depend only on p and K.

Proof. First we will compute the norm of an element in Z.

$$\|(w_n^{\frac{2}{p-2}} T_n y)\|_{p,2,(w_n)} = \max\{(\sum w_n^{\frac{2p}{p-2}} \|T_n y\|_p^p)^{1/p}, (\sum w_n^{\frac{4}{p-2}} \|T_n y\|_2^2 w_n^2)^{1/2}\}$$

$$\sim \max\{(\sum w_n^{\frac{2p}{p-2}} \|y\|_p^p)^{1/p}, (\sum w_n^{\frac{2p}{p-2}} \|y\|_2^2)^{1/2}\}$$

$$= (\sum w_n^{\frac{2p}{p-2}})^{1/p} \max\{\|y\|_p, \|y\|_2 (\sum w_n^{\frac{2p}{p-2}})^{\frac{p-2}{2p}}\}.$$

Thus the map $(\sum w_n^{\frac{2p}{p-2}})^{1/p} y \to (w_n^{\frac{2}{p-2}} T_n y)$ defines an isomorphism from Y in the weighted norm onto Z. Next we define an operator P by

$$P(x_n) = ((\sum w_j^{\frac{2p}{p-2}})^{-1} w_n^{\frac{2}{p-2}} T_n [\sum w_j^{\frac{2(p-1)}{p-2}} T_j^{-1} P_j x_j]),$$

for every sequence $(x_n) \in (\sum X_n)_{p,2,(w_n)}$. Clearly P maps into Z and $Pz = z$ for all $z \in Z$.

It remains to check that P is $(p,2)-$bounded.

$$\|P(x_n)\| \sim (\sum w_n^{\frac{2p}{p-2}})^{\frac{1-p}{p}} \max\{\|y\|_p, \|y\|_2 (\sum w_n^{\frac{2p}{p-2}})^{\frac{p-2}{2p}}\}$$

where $y = \sum w_j^{\frac{2(p-1)}{p-2}} T_j^{-1} P_j x_j$. Since by Hölder's Inequality with exponent pairs $(p, p/(p-1))$ and $(2,2)$,

$$\|y\|_p \leq \sum w_j^{\frac{2(p-1)}{p-2}} \|T_j^{-1}\|_p \|P_j\|_p \|x_j\|_p \leq (\sum w_j^{\frac{2p}{p-2}})^{\frac{p-1}{p}} K^2 (\sum \|x_j\|_p^p)^{1/p}$$

and

$$\|y\|_2 \leq \sum w_j^{\frac{p}{p-2}} w_j \|T_j^{-1}\|_2 \|P_j\|_2 \|x_j\|_2 \leq (\sum w_j^{\frac{2p}{p-2}})^{1/2} K^2 (\sum \|x_j\|_2^2 w_j^2)^{1/2}.$$

Then

$$\|P(x_n)\| \leq K^2 \max\{(\sum \|x_n\|_p^p)^{1/p}, (\sum \|x_n\|_2^2 w_j^2)^{1/2}.$$

$\square$

The next lemma shows that $(p,2)-$sums of subspaces of L_p behave in a reasonable way with respect to $(p,2)-$isomorphisms of their summands. Below we, as usual, consider an ordinal as the set of ordinals less than the given ordinal.

LEMMA 2.7. *Suppose that for some $\alpha < \omega_1$, $\{X_\beta : \beta < \alpha\}$ is a family of subspaces of $L_p(\Omega, \mu)$ for some probability measure μ, and there is a constant K such that for every $\gamma < \beta < \alpha$ there is a $(p,2)-$ continuous projection $Q_{\gamma,\beta}$ from X_β onto a subspace Y_γ such that $\|Q_{\gamma,\beta}\|_{p,2} \leq K$ and Y_γ is K $(p,2)-$ isomorphic to X_γ, in the natural $(p,2)-$ norm. Let (w_j) be a sequence contained in $(0,1]$ and let ϕ be a map from $\mathbb{N}$ into α such that there is an infinite subset M of $\mathbb{N}$ with $\lim_{n \in M} \phi(n) = \alpha$ and for every $\epsilon > 0$,*

$$\sum_{\substack{m \in M \\ w_m < \epsilon}} w_m^{2p/(p-2)} = \infty.$$

Then there is a constant C depending on K and p such that $(\sum_{n=1}^\infty X_{\phi(n)})_{p,2,(w_n)}$ is C $(p,2)-$ isomorphic to $(\sum_{n=1}^\infty X_{\alpha_n})_{p,2,(w_n')}$, where (w_n') is any sequence in $(0,1]$ satisfying (*) and $\lim_{n \to \infty} \alpha_n = \alpha$.

Proof. We proceed in the same way that Rosenthal did in the proof of [R,Theorem 13]. First we single out a special $(p,2)-$ sum. Let $\psi : \mathbb{N} \times \mathbb{N} \to \alpha$ be a function such that for any n, $\psi(n,j)$ is independent of j and $(\psi(n,1))_{n \in \mathbb{N}}$ is an enumeration of α. Let $Z = (\sum_{n,j} X_{\psi(n,j)})_{p,2,(w_{n,j}')}$, where $w_{n,j}' = w_n'$ for all n,j. Clearly if (w_n') satisfies (*) then we can find disjoint finite subsets of $\mathbb{N}$, $N_{n,j}$, such that $\alpha_k > \psi(n,j)$ for all $k \in N_{n,j}$ and $w_j' > (\sum_{k \in N_{n,j}} w_k^{\frac{2p}{p-2}})^{\frac{p-2}{2p}} > w_j'/2$ for all $n,j \in \mathbb{N}$. It follows from Lemma 2.6 that $X_{\psi(n,j)}$ is $(p,2)$ isomorphic to a $(p,2)-$ complemented subspace of $(\sum_{n \in N_{n,j}} X_{\alpha_n})_{p,2,(w_n')}$. An application of Lemma 2.5 gives us that Z is $(p,2)-$ isomorphic to a $(p,2)-$ complemented subspace of $(\sum_{n=1}^\infty X_{\alpha_n})_{p,2,(w_n')}$. Similarly, Z is $(p,2)-$ isomorphic to a $(p,2)-$ complemented subspace of $(\sum_{n=1}^\infty X_{\phi(n)})_{p,2,(w_n)}$. Another application of this argument shows that Z contains $(p,2)-$ complemented $(p,2)-$ isomorphs of $(\sum_{n=1}^\infty X_{\alpha_n})_{p,2,(w_n')}$ and $(\sum_{n=1}^\infty X_{\phi(n)})_{p,2,(w_n)}$.

Finally we note that $(\sum_{n=1}^\infty Z)_{p,2,(1)}$ is $(p,2)-$ isomorphic to Z and thus by the decomposition method (See [R, Proposition 11].) Z is $(p,2)-$ isomorphic to $(\sum_{n=1}^\infty X_{\alpha_n})_{p,2,(w_n')}$ and $(\sum_{n=1}^\infty X_{\phi(n)})_{p,2,(w_n)}$. $\square$

Proof of Theorem 2.4. It is sufficient to prove this for $j = 1$ and then we may conclude by applying Lemma 2.5 and induction. We follow the pattern of the argument given above for the case $\alpha = 1$.

It follows from Lemma 2.1 that $R_p^{\alpha+\omega}$ is $(p,2)-$ isomorphic to

$$\left(\sum R_{p,0}^{\alpha+k}\right)_{p,2,(1)} \oplus_{p,2} L_p^0.$$

Let us compare this with $(\sum R_p^{\alpha+k})_{p,2,(1)}$. A typical element of $R_p^{\alpha+k} = L_p^k \otimes R_p^\alpha$ is of the form

$$f = \sum_m (\sum_{j=0}^{2^k} a_j^k 1_{I_j^k}) \otimes y_m^k = \sum_{j=0}^{2^k} 1_{I_j^k} \otimes (\sum_m a_{m,j}^k y_m^k) = \sum_{j=0}^{2^k} 1_{I_j^k} \otimes h_j^k,$$

where $y_m^k \in R_p^\alpha$ for all k, m and $h_j^k = \sum_m a_{m,j}^k y_m^k$. Thus a typical element of $(\sum R_p^{\alpha+k})_{p,2,(1)}$ is $f = (\sum_{j=0}^{2^k} 1_{I_j^k} \otimes h_j^k)_k$ and

$$\|f\|_{p,2,(1)} = \max\{(\sum_{k=1}^\infty \sum_{j=0}^{2^k-1} \|h_j^k\|_p^p 2^{-k})^{1/p}, (\sum_{k=1}^\infty \sum_{j=0}^{2^k-1} \|h_j^k\|_2^2 2^{-k})^{1/2}\}.$$

If we define $g_j^k = h_j^k 2^{-k/p}$ and $w_j^k = 2^{-k(p-2)/(2p)}$ for each j, k, then

$$\|f\|_{p,2} = \max\{(\sum_{k=1}^\infty \sum_{j=0}^{2^k-1} \|g_j^k\|_p^p)^{1/p}, (\sum_{k=1}^\infty \sum_{j=0}^{2^k-1} \|g_j^k\|_2^2 (w_j^k)^2)^{1/2}\}.$$

Thus $(\sum R_p^{\alpha+k})_{p,2,(1)}$ is isometric to $(\sum R_p^\alpha)_{p,(w_k^n)}$. Now just as in the case $\alpha = 1$ for each n we let $1_n = (x_k)_k$ where $x_n = 1_{[0,1]}$ and $x_k = 0$ for $k \neq n$. Observe that the operator $P(f_n) = ((\int f_n) 1_n)$ is a $(p,2)-$norm 1 projection onto $[1_n : n \in \mathbb{N}]$ with kernel $(\sum R_{p,0}^{\alpha+k})_{p,2,(1)}$. $[1_n : n \in \mathbb{N}]$ is $(p,2)-$isometric to $X_{p,2,(1)}$ which is isomorphic to ℓ_2. Moreover, X_p is $(p,2)-$isomorphic to a $(p,2)-$complemented subspace of $(\sum R_{p,0}^{\alpha+k})_{p,2,(1)}$. Indeed, fix $f \in L_{p,0}^2$ and observe that $L_p^k \otimes f \subset R_p^{\alpha+k+2} = L_p^{k+2} \otimes R_p^\alpha$ and is $(p,2)-$complemented for all k. Thus, just as in the computation preceding Theorem 2.4, $(\sum R_{p,0}^{\alpha+k})_{p,2,(1)} \oplus_{p,2} X_{p,2,(1)}$ is $(p,2)-$isomorphic to $(\sum R_p^{\alpha+k})_{p,2,(1)}$. Hence $(\sum R_{p,0}^{\alpha+k})_{p,2,(1)}$ and $(\sum R_p^{\alpha+k})_{p,2,(1)}$ are $(p,2)-$isomorphic.

It follows that $(\sum R_p^{\alpha+k})_I$ is $(p,2)-$isomorphic to $(\sum R_p^\alpha)_{p,(w_k^n)}$. $\square$

Remark 2.8. By Lemma 2.7 we can replace the weights $(w_{k,n})$ in Theorem 2.4 by any other sequence satisfying (*).

If we let $\alpha = 0$ and write out explicitly the norms in Theorem 2.4, we get the following.

COROLLARY 2.9. $R_p^{\omega k}$ *is* $(p,2)-$*isomorphic to the space*

$$\{(a_{n_1,n_2,\ldots,n_k}) \in \mathbb{R}^{\mathbb{N}^k} : \|(a_{n_1,\ldots,n_k})\| < \infty\}$$

where

$$\|(a_{n_1,\ldots,n_k})\| = \max_{0 \le j \le k}\left\{\left(\sum_{n_1} \cdots \sum_{n_j}\right.\right.$$
$$\left.\left(\sum_{n_{j+1}} \cdots \sum_{n_k} a(n_1,\ldots,n_k)^2 W^2(\{1,\ldots,j\},n_1,\ldots,n_k)\right)^{\frac{p}{2}}\right)^{\frac{1}{p}}\right\},$$

and $W^2(\{1,\ldots,j\},n_1,\ldots,n_k) = \prod_{i=j+1}^k (w_{n_i}^i)^2$.

Notice that the norm given in Corollary 2.9 and that in Proposition 1.3 are similar in form but there are fewer terms. This results in a lack of symmetry in the formula in Corollary 2.9. We will see much later that this difference is in fact significant.

By Lemma 2.7.

$$\lim_{k \to \infty} w k = w^2 \qquad R_p^{w^2} = \sum_{\alpha < w^2} R_p^\alpha = \sum_k R_p^{wk}$$

COROLLARY 2.10. *For every* $\alpha, \omega \leq \alpha < \omega_1$ *and* $k \in \mathbb{N}$, $R_p^{\alpha+k}$ *is* $(p,2)-$ *isomorphic to* R_p^α.

Proof. By Theorem 2.4 R_p^α is isomorphic to a $(p,2)-$sum of spaces X_β as in Lemma 2.7. From Lemma 2.7 it follows that R_p^α is isomorphic to its square. Clearly $R_p^{\alpha+k}$ is isomorphic to a finite direct sum of copies of R_p^α and thus to R_p^α. $\square$

In the next section we will show that the converse of Corollary 2.10 holds and thus it is actually a characterization. Our last goal in this section is to prove that for each $k \in \mathbb{N}$ the space $R_p^{\omega k}$ is $(p,2)-$isomorphic to a complemented subspace of $\otimes^k X_p$. This will be accomplished by using the fact that $R_p^{\omega k}$ is $(p,2)-$isomorphic to the k-fold $(p,2,(w_n))$-sum of X_p, where (w_n) is any sequence in $[0,1]$ satisfying (*). For sequences which do not satisfy (*) we get results similar to those of Rosenthal. Proposition 2.11 is the analogue of [R, Theorem 13] and the consequences noted in [R, p 283].

PROPOSITION 2.11. *If X is a subspace of $L_p(\mu)$ for some probability measure μ, and if (w_n) is a sequence in $(0,1]$ there are four possible isomorphic types for $Y = (\sum X)_{p,2,(w_n)}$. The classes are determined as follows.*

(1) *If* $\inf w_n > 0$, Y *is* $(p,2)-$*isomorphic to* $(\sum X)_{p,2,(1)}$.

(2) *If* $\sum w_n^{2p/(p-2)} < \infty$, Y *is isomorphic to* $(\sum X)_p$.

(3) *If* $\sum w_n^{2p/(p-2)} = \infty$, $\inf w_n = 0$, *and there exists* $\epsilon > 0$ *such that*

$$\sum_{w_n < \epsilon} w_n^{2p/(p-2)} < \infty,$$

then Y is isomorphic to $(\sum X)_{p,2,(1)} \oplus (\sum X)_p$.

(4) *If*

$$\sum_{w_n < \epsilon} w_n^{2p/(p-2)} = \infty$$

for every $\epsilon > 0$, *then Y is* $(p,2)-$*isomorphic to* $(\sum X)_{p,2,(w_n')}$, *where* $w_n' = n^{-(p-2)/(2p)}$.

Proof. Let (x_n) be a sequence in X. It is easy to see that the third case follows from the first two and that the fourth case is included in Lemma 2.7. Thus we need only compute the first two cases.

If $\inf w_n > 0$, then

$$\sum \|x_n\|_2^2 \geq \sum \|x_n\|_2^2 w_n^2 \geq (\inf w_n^2) \sum \|x_n\|_2^2.$$

If $\sum w_n^{2p/(p-2)} < \infty$ then

$$\sum \|x_n\|_2^2 w_n^2 \leq (\sum \|x_n\|_2^p)^{2/p} (\sum w_n^{2p/(p-2)})^{(p-2)/p}$$
$$\leq (\sum \|x_n\|_p^p)^{2/p} (\sum w_n^{2p/(p-2)})^{(p-2)/p}.$$

$\square$

We will next show that $R_p^{\omega k}$ is isomorphic to a complemented subspace of $\otimes^k X_p$. The proof is inductive, so we first prove a proposition which we can use iteratively.

PROPOSITION 2.12. *Suppose that Y is a subspace of $L_p(\mu)$ for some probability measure μ and that (w_n) and (w'_n) are sequences in $(0,1]$. For each $n, k \in \mathbb{N}$ let $w_{n,k} = w_n$. Then $X_{p,2,(w'_n)} \otimes (\sum Y)_{p,2,(w_{n,k})}$ contains a $(p,2)-$complemented subspace which is $(p,2)-$isomorphic to*

$$\left(\sum \left(\sum Y\right)_{p,2,(w_n)}\right)_{p,2,(w'_n)}.$$

Proof. For each $k \in \mathbb{N}$ let $Y_k = [(y_{n,j}) : y_{n,j} = 0 \text{ if } j \neq k] \subset (\sum Y)_{p,2,(w_{n,k})}$. Clearly Y_k is $(p,2)-$isometric to $(\sum Y)_{p,2,(w_n)}$. Let (x_n) be the standard basis of $X_{p,2,(w'_n)}$ and consider the space $Z = [x_k \otimes y_k : y_k \in Y_k \text{ and } k \in \mathbb{N}]$. In order to compute the norm in the tensor product we must first represent $(\sum Y)_{p,2,(w_{n,k})}$ as a subspace of $L_p(\nu)$, for some probability measure ν. We may assume that X contains only mean 0 functions and thus as in Remark 2.3 we can consider ν as an infinite product measure and each summand of $(\sum Y)_{p,2,(w_{n,k})}$ as a space of functions depending only on the (n,k)-coordinate of the product space. Similarly we realize (x_n) as a sequence of L_p-norm one independent mean 0 random variables on some other probability space. Thus a typical element of Z is of the form $\sum x_k \otimes y_k$ and by Rosenthal's inequality has norm equivalent to

$$\max\{(\sum \|x_k \otimes y_k\|_p^p)^{1/p}, (\sum \|x_k \otimes y_k\|_2^2)^{1/2}\}$$
$$= \max\{(\sum \|x_k\|_p^p \|y_k\|_p^p)^{1/p}, (\sum \|x_k\|_2^2 \|y_k\|_2^2)^{1/2}\}$$
$$= \max\{(\sum \|y_k\|_p^p)^{1/p}, (\sum (w'_n)^2 \|y_k\|_2^2)^{1/2}\}.$$

This is precisely the norm in $(\sum Y_k)_{p,2,(w'_n)}$.

To see that Z is complemented it is sufficient to observe that

$$X_{p,2,(w'_n)} \otimes \left(\sum Y\right)_{p,2,(w_{n,k})}$$

has an unconditional decomposition with summands $[x_n] \otimes Y_k$, $n, k \in \mathbb{N}$. (This follows from Lemma 1.2.) □

COROLLARY 2.13. *Suppose that (w_n) is a sequence in $(0,1]$ which satisfies (*). Then*

$$\otimes^k X_{p,2,(w_n)}$$

contains a $(p,2)-$complemented subspace $(p,2)-$isomorphic to

$$\overbrace{\left(\sum \cdots \left(\sum \mathbb{R}\right)_{p,2,(w_n)} \cdots \right)_{p,2,(w_n)}}^{k \text{ times}}.$$

Consequently, $R_p^{\omega k}$ is $(p,2)-$isomorphic to a $(p,2)-$complemented subspace of $\otimes^k X_p$.

Proof. Because (w_n) satisfies (*), $(\sum \mathbb{R})_{p,2,(w_{n,k})}$ is $(p,2)-$isomorphic to $(\sum \mathbb{R})_{p,2,(w_n)}$. Thus the $k = 2$ case follows from Proposition 2.12. Similarly, because Lemma 2.7 implies that

$$\overbrace{\left(\sum \cdots \left(\sum \mathbb{R}\right)_{p,2,(w_n)} \cdots \right)_{p,2,(w_n)}}^{k \text{ times}}$$

is $(p, 2)-$isomorphic to

$$(\sum \overbrace{(\sum \cdots (\sum \mathbb{R})_{p,2,(w_n)} \cdots)_{p,2,(w_n)}}^{k-1 \text{ times}})_{p,2,(w_{n,k})}.$$

The general case follows by induction. $\square$

ISOMORPHIC CLASSIFICATION OF R_p^α, $\alpha < \omega_1$

To complete the isomorphic classification of the spaces R_p^α, $\alpha < \omega_1$, we will develop an invariant based on the presence of nicely placed copies of l_2, In a certain gross sense the proof is similar to that of the isomorphic classification of the spaces $C(\alpha)$, [BP]. We begin with a few definitions.

DEFINITION 3.1. A sequence of measurable functions (f_n) on $(\Omega, \mathcal{B}, \mu)$ will be said to be conditionally independent if there exists a measurable set A such that $\text{supp}(f_n) \subset A$ for all $n \in \mathbb{N}$ and the restrictions to A, $(f_n|_A)$, are independent for the restricted measure space and normalized measure, $\mu|_A/\mu(A)$. The set A is said to be a conditioning set for (f_n).

Conditionally independent sequences occur naturally in the spaces $R_p^{\alpha+k}$. For example if (x_n) is an independent sequence in R_p^α then for any set B which is measurable with respect to the dyadic σ-algebra $\mathcal{D}_k$, $(1_B \otimes x_n)$ is a conditionally independent sequence in $R_p^{\alpha+k}$.

We will be working with basic sequences in L_q, $1 < q < 2$, which are equivalent to the basis of l_2 which are well complemented in the sense that the orthogonal projection is bounded. It is easy to see [F, Lemma 4.4] that if $[x_n]$ is orthogonally complemented in $L_q(\Omega)$ and B is a measurable set of Ω_1 then $[1_B \otimes x_n]$ is orthogonally complemented in $L_q(\Omega_1 \times \Omega)$.

DEFINITION 3.2. Suppose that (x_n) is a sequence in L_p, $p > 2$, which is equivalent to the basis of l_2. Let

$$W(x_n) = \lim_{n \to \infty} \sup\{\frac{\|x\|_2}{\|x\|_p} : x \in [x_k : k \geq n]\}.$$

If (x_n) is a sequence in L_q, $1 < q < 2$, which is equivalent to the basis of l_2 and has complemented closed span with biorthogonal functionals (x_n^*), then let $W(x_n) = W(x_n^*)$.

This definition for $q < 2$ depends on the choice of biorthogonal functionals. In our applications the projection will be the orthogonal projection and thus the choice will be made automatically. The next lemma illustrates the effect of conditional independence on $W()$.

LEMMA 3.3. *If (x_n) is a conditionally independent sequence of mean 0 functions in L_p, $p > 2$, which is equivalent to the basis of l_2 and A is a conditioning set for (x_n), then*

$$\mu(A)^{(p-2)/2p}/K_p \leq W(x_n) \leq \mu(A)^{(p-2)/2p}.$$

Proof. By Hölder's inequality

$$\|x1_A\|_2^2 \leq \|x^2\|_{p/2}\|1_A\|_{p/(p-2)} = \|x\|_p^2\mu(A)^{(p-2)/p}.$$

This proves the right-hand inequality. For the left-hand we use Rosenthal's inequality for the measure $\nu = \mu|_A/\mu(A)$. For convenience we assume that $\|x_n\|_{L_p(\nu)} = 1$.

$$\|\sum_{n=1}^N x_n\|_p\mu(A)^{-1/p} = \|\sum_{n=1}^N x_n\|_{L_p(\nu)}$$

$$\leq K_p\max\{(\sum_{n=1}^N \|x_n\|_{L_p(\nu)}^p)^{1/p}, (\sum_{n=1}^N \|x_n\|_{L_2(\nu)}^2)^{1/2}\}$$

$$= K_p\max\{N^{1/p}, (\sum_{n=1}^N \|x_n\|_2^2\mu(A)^{-1})^{1/2}\}.$$

Because (x_n) is equivalent to the basis of l_2, there is a constant $c > 0$ such that $\|x_n\|_2 \geq c$ for all n. Therefore

$$(\sum_{n=1}^N \|x_n\|_2^2\mu(A)^{-1})^{1/2} \geq N^{1/2}c/\mu(A)^{1/2} > N^{1/p}$$

for N sufficiently large. For such N, we have

$$\|\sum_{n=1}^N x_n\|_p\mu(A)^{-1/p} \leq K_p(\sum_{n=1}^N \|x_n\|_2^2\mu(A)^{-1})^{1/2} = \frac{K_p}{\lambda}\|\sum_{n=1}^N x_n\|_2/\mu(A)^{1/2}.$$

Thus

$$\mu(A)^{(p-2)/2p}/K_p \leq \|\sum_{n=1}^N x_n\|_2/\|\sum_{n=1}^N x_n\|_p.$$

Because the argument applies $(x_{n+s})_{n=1}^N$ for any $s \in \mathbb{N}$, $\mu(A)^{(p-2)/(2p)}/K_p \leq W(x_n)$. $\square$

We will also need to apply the function $W()$ to certain $(p, 2)-$sums. These spaces will in general be isomorphic to complemented subspaces of L_p but it is convenient to use the same notions in the given norm. In particular in the next lemma we will use an isomorph of D_p, [F, p. 108], $(\sum_{n,k} l_2)_{p,2,(w_{n,k})}$ where $w_{n,k} = (2^{-n})^{(p-2)/2p}$ for all k, n. For $p > 2$, the norm we will use is

$$\|\sum a_{n,k}^j e_{n,k}^j\| =$$
$$\max\{(\sum_{n,k}\sum_j |a_{n,k}^j|^p)^{1/p}, (\sum_{n,k}(\sum_j |a_{n,k}^j|^2)^{p/2})^{1/p}, (\sum_{n,k}\sum_j |a_{n,k}^j|^2 w_{n,k}^2)^{1/2}\}$$

where $e^j_{n,k}$ denotes the sequence which is 0 except at the (j,n,k)-th place where it is 1. For this space

$$\|\sum a^j_{n,k} e^j_{n,k}\|_2 = (\sum_{n,k} \sum_j |a^j_{n,k}|^2 w^2_{n,k})^{1/2}$$

and we will compute $W(x_n)$ for $(x_n) \subset D_{p,(w_{n,k})}$ from the ratio $\|x\|_2/\|x\|$ for $x \in [x_n : n > k]$. For $q < 2$, the norm on D_q is the dual norm to the norm on D_p, where p is the conjugate index to q. In the definition of $W()$ for D_q we will take the biorthogonal functionals from D_p and proceed analogously. Notice that for fixed n,k, $(e^j_{n,k} : j \in \mathbb{N}) \subset D_p, p > 2$ is 1-equivalent to the basis of l_2 and has 1-complemented span and thus so are the dual functionals. Further, for fixed j, $(e^j_{n,k} : n,k \in \mathbb{N})$ is 1-equivalent to the basis of $X_{p,2,(w_{n,k})}$.

The next lemma is a key step in the proof and is analogous to Schechtman's result, [S, Proposition 2], that the natural basis of $(\sum l_s)_{l_r}, 1 < q < r < s \leq 2$, is not equivalent to a sequence of independent random variables in L_q. We use several arguments which are given in that paper and leave the reader to consult [S] for additional detail.

LEMMA 3.4. *If T is a bounded linear map from $(\sum l_2)_{q,2,(w_{n,k})}$ into L_q, $1 < q < 2$, and the range of T is contained in $[x_n]$ where (x_n) is a sequence of independent mean zero random variables which are basic, then there is a normalized block basic sequence $(y^j_{n,k})$ such that $[y^j_{n,k}]$ is a $(q,2)-$ complemented subspace of $(\sum l_2)_{q,2,(w_{n,k})}$, $(y^j_{n,k})$ is equivalent to the standard basis of $(\sum l_2)_{q,2,(w_{n,k})}$, and for each n,k, $W(y^j_{n,k}) = w_{n,k}$, and $\|T(y^j_{n,k})\| < 2^{-n-k}$ for all j.*

Proof. Let $(e^j_{n,k})$ be the natural basis of $(\sum l_2)_{q,2,(w_{n,k})}$ where $W(e^j_{n,k} : j \in \mathbb{N}) = w_{n,k}$. We can assume that $\liminf_{k\to\infty} \liminf_{j\to\infty} \|Te^j_{n,k}\| > 0$ for each n, otherwise simple averaging of $(e^k_{n,j})$ will produce the required sequence. (See end of the proof below where we use a similar argument.) By a diagonal argument and passing to a subsequence of the index j for each n,k we may assume that $(T(e^j_{n,k}))$ is disjointly supported relative to the basic sequence (x_n).

For each n, $(e^j_{n,k})_{k,j}$ is equivalent to the basis of l_2 and thus $(Te^j_{n,k})_{k,j}$ is q-equi-integrable (See [S] for the definition.). By [S, Lemma 4] and the proof of [S, Proposition 2] we can find a subsequence of the k's (which we assume to be the whole sequence for notational convenience) and for each k a subsequence of the j's (which we again assume is the whole sequence) such that $(Te^j_{n,k})_j$ is $(q, 2^{-i})$-equi-distributed and $(Te^j_{n,k})_k$ is $(q, 2^{-t})$-equi-distributed. (The size of i and t are determined by the estimate of symmetry required in the third display below.) However for n fixed and large and t and i sufficiently large the behavior in the domain is quite different. Indeed,

$$\|\sum_{j=1}^M e^j_{n,k}\| = M^{1/2}$$

and

$$\|\sum_{k=1}^M e^j_{n,k}\| = M^{1/q}$$

if $M < w_n^{2q/(q-2)}$. Thus

$$\|T\|M^{1/2} \geq \|T(\sum_{j=1}^{M} e_{n,k}^j)\| \sim \|T(\sum_{k=1}^{M} e_{n,k}^j)\|.$$

Therefore

$$\|T(M^{-1/q}\sum_{k=1}^{M} e_{n,k}^j)\| \leq 2\|T\|M^{(q-2)/2q}.$$

Also note that if M is essentially equal to $(w_m/w_n)^{2q/(2-q)}$ then the 2 and p-norms of the corresponding dual functional are approximately equal to w_m. Now select integers $M_{k,m}$ and $n_{k,m}$ such that $2\|T\|M_{k,m}^{(q-2)/2q} < 2^{-k-m}$ and $M_{k,m} \sim (w_m/w_{n_{k,m}})^{2q/(2-q)}$. (In particular, for a given m we need to choose $n_{k,m}$ so that $(w_m/w_{n_{k,m}})^{2q/(2-q)} > 2^{k+m}\|T\|$. We also assume that approximations improve as $k \to \infty$.)

For $k, m \in \mathbb{N}$ find disjoint sets $K_{k,m}$ of $\mathbb{N}$ of cardinality $M_{k,m}$. (Technically we must actually choose appropriate subsequences of the index set j so that the required estimates above hold for $n = n_{k,m}$ and $M = M_{k,m}$. We also need to ensure that all of the blocks we construct are disjointly supported. All of these are possible by simple diagonalization but are messy notationally, so we leave the details to the reader.) Let $y_{m,k}^j = M_{k,m}^{-1/q}\sum_{s \in K_{k,m}} e_{n_{k,m},s}^j$. Then

$$\|\sum_{s \in K_{k,m}} e_{n_{k,m},s}^j{}^*\|_2 / \|\sum_{s \in K_{k,m}} e_{n_{k,m},s}^j{}^*\|_p \sim M_{k,m}^{1/2}w_{n_{k,m}}/M_{k,m}^{1/p}$$

which converges to w_m as k goes to ∞. Thus $W(y_{m,k}^j, j \in N) \sim w_m$. Therefore $[y_{m,k}^j : j \in N]$ is $(q, 2)$−isomorphic to $X_{q,2,(w_m)}$ (m is fixed so the sequence (w_m) is constant.) and by (duality and) Proposition 0.2 is $(q, 2)$−complemented with a bound independent of k and m. It follows from Lemma 2.5 that $[y_{k,m}^j]$ is $(q, 2)$− complemented. □

Remark 3.5. In the application of Lemma 3.4 and the next one we will actually be working in some R_q^α rather than in sequence norm of $(\sum l_2)_{q,2,(w_{n,k})}$. Therefore we need to make some observations about the $(q, 2)$−norms that occur and the constant $W(y_{k,m}^j)$. First if for each m, k we construct the sequence $(y_{k,m}^j)$ in L_q so that the span is orthogonally complemented in L_q, then the estimates can be made in $L_p = L_q^*$ with the dual functionals. Let us examine the dual situation in L_p more closely.

In [F, Theorem 4.8] it is shown that if for each $n \in \mathbb{N}$, X_n is an orthogonally complemented subspace of $L_p[0, 1]$ (and mean 0) and the norms of the projections do not depend on n, then the space constructed by squeezing the supports onto sets of measure $w_n^{2p/(p-2)}$ by the obvious L_p-isometry, and then placing the resulting spaces on independent coordinates produces a $(p, 2)$−complemented subspace of $L_p(\prod[0, 1])$ which is isomorphic to $(\sum X_n)_{p,2,(w_n)}$. Moreover, the image of X_n in the resulting space is still orthogonally complemented (with the same norm) and the only significant change is that l_2-norms of the elements have been multiplied by a constant, w_n. Second if the spaces X_n are actually the span of mean 0, independent random variables $(x_{n,k})$ with $\|x_{n,k}\|_2/\|x_{n,k}\|_p = c$, then the distance to l_2 and the norm of the orthogonal projection onto $[x_{n,k} : k \in \mathbb{N}]$ is a function of c and the constants in Rosenthal's inequality.

Finally notice that in the proofs of Lemmas 3.4 and 3.6 only simple averages of elements whose dual functionals have the same ratio for p and 2-norms are employed. Therefore the resulting dual functionals are also simple averages. (Of course properly normalized.)

We will actually need to consider a sequence of maps rather than just one and so we will use diagonalization to extend Lemma 3.4 as follows.

LEMMA 3.6. *If (T_r) is a sequence of bounded linear maps from $(\sum l_2)_{q,2,(w_{n,k})}$ into L_q, $1 < q < 2$, and the union of the ranges of $T_r, r \in \mathbb{N}$, is contained in $[x_n]$ where (x_n) is a sequence of independent mean zero random variables which are basic, then there is a normalized block basic sequence $(y_{n,k}^j)$ such that $[y_{n,k}^j]$ is a $(q,2)-$ complemented subspace of $(\sum l_2)_{q,2,(w_{n,k})}$, $(y_{n,k}^j)$ is equivalent to the standard basis of $(\sum l_2)_{q,2,(w_{n,k})}$, and for each n,k, $W(y_{n,k}^j) = w_{n,k}$, and $\|T_r(y_{n,k}^j)\| < 2^{-n-k}$ for all $r < n+k$ and all $j \in \mathbb{N}$.*

Sketch of Proof. For a given m and k only the finitely many operators T_r, such that $r < m + k$, need be considered. In the proof of Lemma 3.4 the choice of M depends only on m and k and the norm of T. For a finite number of operators the requirement that $T_r(e_{n,k}^j)$ be a block subsequence can be achieved simultaneously by passing to subsequences and the estimates

$$\|T_r\|M^{1/2} \geq \|T_r(\sum_{j=1}^{M} e_{n,k}^j)\| \sim \|T_r(\sum_{k=1}^{M} e_{n,k}^j)\|.$$

can be achieved for all $r < m + k$ as well. Thus the proof above generalizes to this case. $\square$

Remark 3.7. The previous result could also have been obtained by applying Lemma 3.4 iteratively and choosing k so that $T_l(y_{n,k}^j)$ is small for all $l < r$ and then using the fact that a sequence of independent random variables has an upper l_q estimate to see that

$$\|T_l(M^{-1/q} \sum_{k=1}^{M} y_{n,k}^j)\| \leq K(\sum_{k=1}^{M} \|T_l(y_{n,k}^j)\|^q)^{1/q}/M^{1/q} \leq K \max \|T_l(y_{n,k}^j)\|.$$

We are now ready for the main result of this section. As might be expected the proof is by transfinite induction.

THEOREM 3.8. *Let $1 < q < 2$. For every $\alpha < \omega_1$, $R_q^{\alpha+\omega}$ is not isomorphic to a subspace of R_q^α.*

Theorem 3.8 is actually a corollary of the following more informative result.

PROPOSITION 3.9. *There are constants C, C' such that if $\omega_1 > \alpha \geq \omega$ and for each $r \in \mathbb{N}$, T_r is a bounded operator from R_q^α into $R_q^{\alpha'}$, $\alpha' + \omega \leq \alpha$, then there is a a normalized basic sequence (y_n) in R_q^α such that (y_n) is equivalent to the standard basis of l_2, $[y_n]$ is C-$(q, 2)-$complemented, $W(y_n) > C'$, and $\|T_r(y_n)\| < 2^{-n}$ for all $r < n$.*

Before we begin the proof of this let us formalize a consequence of the statement above that we will use in the induction.

LEMMA 3.10. *Let X be a subspace of L_q which is complemented by the orthogonal projection and let (x_n) be an unconditional basis for X with dual functionals $(x_n^*) \subset L_p$ such that (x_n^*) is an orthogonal sequence in L_2. Suppose that (X_j) is a sequence of subspaces of L_q such that for any k if (T_r) is a sequence of operators from X into $(\sum_{j=1}^k {}'X_j)_I$, then there is a a normalized basic sequence (y_n) in X such that (y_n) is equivalent to the standard basis of l_2, $[y_n]$ is C-$(q, 2)-$complemented by the orthogonal projection, $W(y_n) > C'$, and $\|T_r(y_n)\| < 2^{-n}$ for all $r < n$. Let P_l denote the projection from $(\sum_{j=1}^\infty {}'X_j)_I$ onto $(\sum_{j=1}^l {}'X_j)_I$. Then for any $\epsilon > 0$, if (T_r) is a sequence of operators from X into $(\sum_{j=1}^\infty X_j)_I$, then there is a a normalized basic sequence (y_n) in X such that (y_n) is equivalent to the standard basis of l_2, $[y_n]$ is $K(C+\epsilon)$-$(q, 2)-$complemented, $W(y_n) > (C'-\epsilon)/K$, where K is the constant in the upper l_2-estimate for (x_n^*), and $\|P_l T_r(y_n)\| < 2^{-n}$ for all $l, r < n$.*

Proof. The proof is a gliding hump argument. Indeed, for each l we use the hypothesis for the sequence of operators $(P_j T_r)_{r=1, j \leq l}^\infty$ to get a sequence (y_n^l) as above. Let $y_1 = y_1^1$, and assume that y_j has been chosen for $j < m$. For n large y_n^m satisfies $\|P_l T_r(y_n^m)\| < 2^{-m}$ for all $l, r < m$. We know that $[y_n^m]$ is complemented. Let (y_n^{m*}) be the dual functionals in X^*. By choosing n large enough we can assume that the support of y_n^{m*} relative to the dual basis (x_n^*) of X^* is disjoint from the supports of y_j^* for $j < m$. By construction $\|y_n^{m*}\|_2 / \|y_n^{m*}\|_p > C'$, for all n, m. Therefore if we choose $y_m = y_{n(m)}^m$, so that $(y_{n(m)}^{m*})$ is an orthogonal sequence then (y_m) will be equivalent to the basis of l_2 and have orthogonally complemented closed span. $\square$

Remark 3.11. The proof of Lemma 3.10 does not really require that X have an unconditional basis, but that it has an unconditional F.D.D. so that the copies of l_2 are taken from different summands. In this way selecting elements from the different l_2 bases will still produce a copy of l_2. Moreover in the application of Lemma 3.10 below, the elements (y_n) will actually be independent and mean 0, so that the constant K in the lemma can be replaced by 1.

Remark 3.12. In [BRS] it is shown that there is a certain isomorph of L_p, $X_{\mathcal{D}}^p$, so that for every $\alpha < \omega_1$, R_p^α is (isometric to a) contractively complemented in $X_{\mathcal{D}}^p$. Also the projection is the orthogonal projection. Moreover, if $\alpha < \beta < \omega_1$, R_p^α is contractively complemented by the orthogonal projection in R_p^β.

Proof of Proposition 3.9. We will use induction on α. Given $\epsilon > 0$, the constants C, C' can be chosen to be $(1+\epsilon)C_p$ and $1-\epsilon$, respectively, where C_p is the $(q, 2)-$norm of the orthogonal projection onto the span of the Rademachers in L_p. At each stage of the induction we actually will produce a badly preserved copy of $(\sum l_2)_{p,2,(w_{n,k})}$.

If $\alpha = \omega$, $R_p^{\alpha'}$ is finite dimensional and the result is obvious. Note also that we may assume that α is a limit ordinal since for $n \in \mathbb{N}$, $R_q^{\alpha+n}$ contains a contractively complemented subspace isometric to R_q^α and $\alpha' + \omega \leq \alpha + n$ implies that $\alpha' + \omega \leq \alpha$. Similarly we may asssume that $\alpha = \alpha' + \omega$, because if $\alpha > \alpha' + \omega$ then R_q^α contains a contractively complemented subspace isometric to $R_q^{\alpha'+\omega}$ and the inductive hypothesis for $\alpha' + \omega$ gives the result.

Suppose $\alpha = \omega \cdot 2$. Because R_q^ω is isomorphic to X_q, we can assume that the union of the ranges of (T_r) is in the span of a sequence of mean zero independent random variables. $R_q^{\omega \cdot 2} = (\sum R_q^{\omega+n})_I$ contains a well $(q,2)-$complemented $((q,2)-$isomorphic) copy of $(\sum l_2)_{q,2,(w_{n,k})}$ since R_q^ω contains well $(q,2)-$complemented copies of $X_{q,2,\delta}$, for any $0 < \delta \leq 1$. In particular, in order to avoid any problems with constants, for each n, k we can take a sequence of independent mean zero Bernoulli random variables and find an (almost) isometric copy of their span in $R_q^{\omega+m(n,k)}$ with the weight $W(y_{n,k}) = w_{n,k}$. Applying Lemma 3.6 to this copy of $(\sum l_2)_{q,2,(w_{n,k})}$ yields the required sequence (y_n). Note that the constants which occur here do not depend on passing through the isomorphisms, because as noted in Remark 3.5 we actually get a sequence (y_n) of independent mean 0 random variables which have closed span complemented by the orthogonal projection from L_q. The $(q,2)-$norm of this projection is the same as the $(q,2)-$norm of the projection onto a similar block in X_q and the isomorphism to l_2 is similarly controlled. At one point in the argument we need the slightly stronger statement that the projections are in fact the orthogonal projections in order to apply Lemma 3.10. Technically our inductive hypothesis should be strengthened to include this information.

Now assume that the result holds for all $\beta < \alpha$. R_q^α is equal to $(\sum_{\beta<\alpha} R_q^\beta)_I$. Let $\beta_n = \alpha' + n$. We only make use of the summands $R_q^{\alpha'+n}$ for $n \in \mathbb{N}$ in $R_q^{\alpha'+\omega}$. Thus for our purposes we can consider $(\sum_n' R_q^{\beta_n})_I$ in place of R_q^α. Also we can replace $R_q^{\alpha'}$ by the (codimension one) isomorphic space $(\sum' R_q^{\beta_l'})_I$, where (β_l') is an enumeration of $\{\beta < \alpha'\}$. For each $n \in N$ let P_n be the orthogonal projection from $(\sum_l' R_q^{\beta_l'})_I$ onto $(\sum_{s=1}^n {}' R_q^{\beta_s'})_I$. (This is in fact just conditional expectation.) Consider the sequence of operators $(T_t')_t = (P_n T_r)_{n,r}$ in some order.

The proof divides into two cases depending on the nature of α'. First we consider the case $\alpha' = \gamma + \omega$, for some $\gamma \geq \omega$. Observe that for any n, $R_q^{\beta_n'}$ is isomorphic to a subspace of R_q^γ and therefore, for each n, $(\sum_{s=1}^n {}' R_q^{\beta_s'})_I$ is also isomorphic to a subspace of R_q^γ. Therefore the inductive hypothesis applies to β_n for all n and thus also for maps from $R_q^{\alpha'+n}$ into $(\sum_{s=1}^n {}' R_q^{\beta_s'})_I$. ($R_q^{\alpha+n}$ contains a well complemented isometric copy of $R_q^{\beta_s+\omega}$ for all $s \in \mathbb{N}$.)

Let $\phi : \mathbb{N} \times \mathbb{N}$ into $\mathbb{N}$ be an injection such that $\phi(n,k) > n$. For each n, k let $X_{n,k}$ be an isometric (in L_q norm) copy of $R_q^{\alpha'}$ in $R_q^{\beta_{\phi(n,k)}}$ which is contractively complemented but is supported on a set of measure $w_{n,k}^{2p/(p-2)}$, i.e., the norm in L_2 has been multiplied by $w_{n,k}$. Therefore for each $n \in \mathbb{N}$ there is a normalized sequence $(z_{n,k}^j)_j$ in $X_{n,k}$ such that $(z_{n,k}^j)_j$ is equivalent to the standard basis of l_2, $[z_{n,k}^j : j \in \mathbb{N}]$ is C-$(q,2)$−complemented, $w_{n,k} \geq W((z_{n,k}^j)_j) > C'w_{n,k}$, and $T_r'(z_{n,k}^j) < 2^{-j}$ for all $r < j$. By a diagonalization argument we can find $J_{n,k} \subset \mathbb{N}$ for all n, k such that $(T(z_{n,k}^j))_{j \in J_{n,k}, n, k}$ is a block of the independent sum $(\sum' R_q^{\beta_l'})_I$. It follows that $[z_{n,k}^j : j \in J_{n,k}, n, k]$ is $(q,2)$−isomorphic to $(\sum l_2)_{q,2,(w_{n,k})}$. Moreover, because of the diagonalization step, the range of T_r is contained in the span of a sequence of independent random variables. Therefore by Lemma 3.6 there is a a normalized block basic sequence $(y_{n,k}^j)$ such that $[y_{n,k}^j]$ is a $(q,2)$−complemented subspace of $[z_{n,k}^j : j \in J_{n,k}, n, k]$, $(y_{n,k}^j)$ is equivalent to the standard basis of $(\sum l_2)_{q,2,(w_{n,k})}$, for each n, k, $W(y_{n,k}^j) = w_{n,k}$, and $\|T_r(y_{n,k}^j)\| < 2^{-n-k}$ for all $r < n + k$ and all $j \in \mathbb{N}$. The subsequence $(y_{1,k}^1)$ meets our requirements.

If α' is not of the form $\gamma + \omega$, then we need to apply Lemma 3.10 to overcome a small difficulty and then proceed similarly to the previous case. As before $R_q^{\alpha'}$ is isomorphic to $(\sum_n' R_q^{\beta_n'})_I$ where (β_n') is an enumeration of the ordinals less than α'. However these are not all isomorphic to subspaces of R_q^γ for some $\gamma < \alpha'$. However if we consider the maps (T_r) from $X_{n,k}$ as above, we have a maps from an isometric copy of $R_q^{\alpha'}$ with its l_2-norm multiplied by $w_{n,k}$ into $(\sum_n' R_q^{\beta_n'})_I$. By the inductive hypothesis a sequence of maps from $R_q^{\alpha'}$ into $(\sum_{n \leq m}' R_q^{\beta_n'})_I$ will satisfy the hypothesis of Lemma 3.10. Therefore by Lemma 3.10 we can find for each n, k a normalized basic sequence $(z_{n,k}^j)$ in $X_{n,k}$ such that $(z_{n,k}^j)$ is equivalent to the standard basis of l_2, $[z_{n,k}^j]$ is C-$(q,2)$−complemented, $w_{n,k} \geq W(z_{n,k}^j) > C'w_{n,k}$, and $T_r'(z_{n,k}^j) < 2^{-j}$ for all $r < j$. At this point the proof continues as in the previous case. $\square$

COROLLARY 3.13. *Suppose that* $\omega \leq \alpha < \alpha' < \omega_1$. *Then* R_p^α *is isomorphic to* $R_p^{\alpha'}$ *if and only if* $\alpha + \omega > \alpha'$.

Proof. One direction is Corollary 2.10 and the other is immediate from Proposition 3.9. $\square$

Remark 3.14. By modifying Schechtman's proof that $\otimes^{k-1}X_q$ does not contain

$$\left(\sum\left(\sum\cdots\left(\sum \ell_{r_1}\right)_{r_2}\cdots\right)_{r_{k-1}}\right)_{r_k},$$

to use D_q as in the argument above, we can also show that for each $k \in \mathbb{N}$, $R_q^{\omega^k}$ is not isomorphic to a subspace of $\otimes^{k-1}X_q$. The point is that the first part of Schechtman's argument uses the induction hypothesis to locate a sequence of independent random variables which contain a copy of $(\sum \ell_{r_{k-1}})_{r_k}$. Our argument needs a sequence of independent random variables which contains a copy of D_q.

CHAPTER 4

ISOMORPHISM FROM $X_p \otimes X_p$ INTO $(p, 2)-$SUMS

In this section we begin to investigate isomorphisms from $X_p \otimes X_p$ into subspaces of L_p. Because, for $p > 2$, $X_p \otimes X_p$ is isomorphic to a subspace of $(\sum l_2)_p$ and $(\sum l_2)_p$ is isomorphic to a (complemented) subspace of $X_p \otimes X_p$, one cannot say too much. However there are some restrictions which are related to the l_2 structure which reveal themselves. In later sections we will consider complemented embeddings and obtain stronger results.

In this section we will be working with X_p as a sequence space. Thus if (e_n) is the usual basis of $X_{p,(w_n)}$, then we will write the norm as

$$\| \sum a_n e_n \| = \max\{\| \sum a_n e_n \|_2, \| \sum a_n e_n \|_p\}$$

where

$$\| \sum a_n e_n \|_2 = (\sum a_n^2 w_n^2)^{1/2} \text{ and } \| \sum a_n e_n \|_p = (\sum a_n^p)^{1/p}.$$

Our first result is about X_p itself. (See [A2] for related results.) Notice that it says that any operator from X_p into L_p acts like an L_2 bounded operator on a large part of the basis.

LEMMA 4.1. *Suppose that (z_m) is a normalized standard basis of X_p and T is an isomorphism of X_p into L_p. If $B = \{m : \|T z_m\|_2 > \|T\|\|z_m\|_2\}$ then*

$$\sum_{m \in B} \|z_m\|_2^{2p/(p-2)} \leq 1.$$

Proof. Let F be a finite subset of B and let $y = \sum_{m \in F} \|z_m\|_2^{2/(p-2)} z_m$. Then

$$\|y\|_2 = \left(\sum_{m \in F} \|z_m\|_2^{2p/(p-2)} \right)^{1/2}$$

and

$$\|y\|_p = \left(\sum_{m \in F} \|z_m\|_2^{2p/(p-2)} \right)^{1/p}$$

Therefore

$$\|T\| \left(\sum_{m \in F} \|z_m\|_2^{2p/(p-2)} \right)^{1/2} < \left(\sum_{m \in F} \|z_m\|_2^{4/(p-2)} \|Tz_m\|_2^2 \right)^{1/2}$$

$$= \left(\int \|T \sum_{m \in F} \|z_m\|_2^{2/(p-2)} r_m(t) z_m\|_2^2 dt \right)^{1/2}$$

$$\leq \left(\int \|T \sum_{m \in F} \|z_m\|_2^{2/(p-2)} r_m(t) z_m\|_p^p dt \right)^{1/p}$$

$$\leq \|T\| \max\{\|y\|_p, \|y\|_2\}.$$

If $\sum_{m \in F} \|z_m\|_2^{2p/(p-2)} > 1$, then $\|y\|_2 > \|y\|_p$, $\max\{\|y\|_p, \|y\|_2\} = \|y\|_2$, and we have a contradiction. $\square$

In the next result we obtain a "type 2 inequality" for the restriction of an operator on $X_p \otimes X_p$ to a large subspace. In $X_p \otimes X_p$ as a sequence space there are four norms which appear in the expression in Proposition 1.3. Below we need only one of them. If (e_n) and (e_n') are copies of the natural basis of $X_{p,(w_n)}$, we will use the notation

$$\|\sum_{n,m} a_{n,m} e_n \otimes e_m'\|_2 = (\sum_{n,m} a_{n,m}^2 w_n^2 w_m^2)^{1/2}$$

for this ℓ_2-norm. In the statement of Proposition 4.2, we assume that the sequence (w_n) contains each value infinitely often, so that $X_{p,(w_n)} = X_{p,(w_{m,k})}$ with $w_{m,k} = w_m'$ for all k.

PROPOSITION 4.2. *If T is an isomorphism from $X_p \otimes X_p$ into L_p and (e_n) and (e_n') are copies of the natural basis of $X_{p,(w_n)}$, then there are complemented subspaces, Y and Z, of X_p each isomorphic to X_p, with standard X_p bases, (y_m) and (z_m), respectively, which are subsequences of (e_n) and (e_n'), respectively, such that $\|T(y_m \otimes z_k)\|_2 \leq \|T\|\|y_m \otimes z_k\|_2$, for all m and k. Consequently, for all $\sum a_{m,k} y_m \otimes z_k \in Y \otimes Z$,*

$$\left(\int \|T \sum a_{m,k} r_{m,k}(t) y_m \otimes z_k\|_2^2 dt \right)^{1/2} \leq \|T\|\|\sum a_{m,k} y_m \otimes z_k\|_2$$

where $(r_{m,k})$ is a doubly indexed set of Rademacher functions.

Proof. Let $(u_{m,n})$ and $(v_{m,n})$ be bases of X_p such that $\|u_{m,n}\|_2 = \|v_{m,n}\|_2 = w_m$ for all m, n and $w_m \downarrow 0$. By the previous lemma for each m and n we have that if

$$B_{m,n} = \{(k,j) : \|T\|\|u_{m,n} \otimes v_{k,j}\|_2 < \|Tu_{m,n} \otimes v_{k,j}\|_2\}$$

and

$$C_{m,n} = \{(k,j) : \|T\|\|u_{k,j} \otimes v_{m,n}\|_2 < \|Tu_{k,j} \otimes v_{m,n}\|_2\}$$

then

$$\sum_{(k,j)\in B_{m,n}} \|u_{m,n}\otimes v_{k,j}\|_2^{2p/(p-2)} \le 1 \quad \text{and} \quad \sum_{(k,j)\in C_{m,n}} \|u_{k,j}\otimes v_{m,n}\|_2^{2p/(p-2)} \le 1.$$

Also note that for all k, j, m, n, $(k,j) \notin B_{m,n}$ if and only if $(m,n) \notin C_{k,j}$.

To get the subspaces Y and Z we will inductively choose disjoint finite subsets F_s and G_s of $\mathbb{N} \times \mathbb{N}$ and infinite subsets M_s, N_s of $\mathbb{N} \times \mathbb{N}$ such that for all $s \in \mathbb{N}$ and for all $(m,n) \in F_s, (k,j) \in G_s$

$$M_{s+1} \subset M_s \text{ and } N_{s+1} \subset N_s,$$

$$\|Tu_{m,n}\otimes v_{k,j}\|_2 \le \|T\|\|u_{m,n}\otimes v_{k,j}\|_2,$$

$$M_s \supset G_t \text{ for } t \le s \text{ and } N_s \supset F_t \text{ for } t \le s,$$

$$|N_s \cap \{(k,r) : r \in \mathbb{N}\}| = \infty \text{ and } |M_s \cap \{(k,r) : r \in \mathbb{N}\}| = \infty \qquad \text{for all } k.$$

Let $F_1 = \{(1,1)\}, M_1 = \mathbb{N} \times \mathbb{N} \setminus B_{1,1}, G_1 = \{(1,j_1)\}$ for some $(1,j_1) \in M_1$ and $N_1 = \mathbb{N} \times \mathbb{N} \setminus C_{1,j_1}$. Note that $(1,1) \notin C_{1,j_1}$ because $(1,j_1) \notin B_{1,1}$, and thus $F_1 \subset N_1$.

Now suppose that we have defined F_s, G_s, M_s and N_s. Let $F_{s+1} \subset N_s \setminus \cup_{t\le s} F_t$ such that F_{s+1} is finite and $F_{s+1} \cap \{k\} \times \mathbb{N} \ne \emptyset$ for $k = 1, 2, \ldots, s+1$. Define $M_{s+1} = M_s \setminus \cup_{(m,n)\in F_{s+1}} B_{m,n}$. Observe that if $t \le s$ then $G_t \cap \cup_{(m,n)\in F_{s+1}} B_{m,n} = \emptyset$ because $F_{s+1} \cap \cup_{(k,j)\in G_t} C_{k,j} = \emptyset$. Choose a finite set $G_{s+1} \subset M_{s+1} \setminus \cup_{t\le s} G_t$ such that $G_{s+1} \cap \{k\} \times \mathbb{N} \ne \emptyset$ for $k = 1, 2, \ldots, s+1$. Define $N_{s+1} = N_s \setminus \cup_{(k,j)\in G_{s+1}} C_{k,j}$. Note that for $t \le s+1$ and $(k,j) \in G_{s+1}$, $F_t \cap C_{k,j} = \emptyset$ because $(k,j) \notin B_{m,n}$ for all $(m,n) \in \cup_{t\le s+1} F_t$.

This completes the induction step. Because the sets F_s are disjoint and $F_s \cap \{k\} \times \mathbb{N} \ne \emptyset$ for all $s \ge k$, $\cup F_s \cap \{k\} \times \mathbb{N}$ is infinite for all k. Similarly the same is true for $\cup G_s$. Thus

$$Y = [u_{m,n} : m, n \in \cup F_s] \text{ and } Z = [v_{m,n} : m, n \in \cup G_s]$$

are isomorphic to X_p. We have that $\|Tu_{m,n}\otimes v_{k,j}\|_2 \le \|T\|\|u_{m,n}\otimes v_{k,j}\|_2$ for all $(m,n) \in \cup F_s, (k,j) \in \cup G_s$ because $(m,n) \in F_s, (k,j) \in G_t, s \le t$, implies that $G_t \subset M_s \subset \mathbb{N} \times \mathbb{N} \setminus B_{m,n}$ and if $t < s$ then $F_s \subset N_t \subset \mathbb{N} \times \mathbb{N} \setminus C_{k,j}$.

The last statement of the conclusion follows from the fact that $(r_{m,k}y_m \otimes z_k)$ is an orthogonal sequence and hence

$$\int \|T\sum a_{m,k}r_{m,k}y_m \otimes z_k\|_2^2 = \sum |a_{m,k}|^2 \|Ty_m \otimes z_k\|_2^2$$

$$\le \|T\|^2 \sum |a_{m,k}|^2 \|y_m \otimes z_k\|_2^2$$

$$= \|T\|^2 \|\sum a_{m,k}y_m \otimes z_k\|_2^2.$$

$\square$

Remark 4.3. If it were possible to pass to subsequences of the bases $(y_m), (z_k)$ for which each still spanned X_p and the image $(Ty_m \otimes z_k)$ was unconditional then the average over signs could be removed from the conclusion of Proposition 4.2. Unfortunately we do not know if this is possible.

In view of Proposition 4.2 we can usually assume that we have passed to the subspaces Y and Z and thus for the given bases of X_p,

(T2) $$\|T(x_i \otimes y_j)\|_2 \leq \|T\|\|x_i \otimes y_j\|_2,$$

for all $i, j \in \mathbb{N}$.

The next result shows that diagonal blocks of the basis of $X_p \otimes X_p$ must be mapped by a bounded operator from $X_p \otimes X_p$ into a $(p, 2)-$sum so that the norm $\| \cdot \|_2$ is well controlled. In the computation we will use some of the other norms from the formula in Proposition 1.3, so we introduce special notation for them. For each $i \in \mathbb{N}$, let

$$\mathcal{R}_i \sum_{n,m} a_{n,m} e_n \otimes e_m' = \sum_m a_{i,m} e_m'$$

and

$$\mathcal{C}_i \sum_{n,m} a_{n,m} e_n \otimes e_m' = \sum_n a_{n,i} e_n.$$

Then define

$$\|\sum_{n,m} a_{n,m} e_n \otimes e_m'\|_R = (\sum_i \|\mathcal{R}_i \sum_{n,m} a_{n,m} e_n \otimes e_m'\|_2^p)^{1/p} = (\sum_i (\sum_m a_{i,m}^2 w_m^2)^{p/2})^{1/p}$$

and

$$\|\sum_{n,m} a_{n,m} e_n \otimes e_m'\|_C = (\sum_i \|\mathcal{C}_i \sum_{n,m} a_{n,m} e_n \otimes e_m'\|_2^p)^{1/p} = (\sum_i (\sum_n a_{n,i}^2 w_n^2)^{p/2})^{1/p}.$$

Finally, define

$$\|\sum_{n,m} a_{n,m} e_n \otimes e_m'\|_p = (\sum_{n,m} a_{n,m}^p)^{1/p}.$$

Thus if $z \in X_p \otimes X_p$, we have

$$\|z\| = \max\{\|z\|_2, \|z\|_R, \|z\|_C, \|z\|_p\}.$$

LEMMA 4.4. *If (Y_n) is a sequence of subspaces of $L_p[0,1]$ with $C-$unconditional bases and $T : X_{p,w} \otimes X_{p,w} \to (\sum Y_n)_{p,2}$ is an isomorphism and (z_n) is a sequence of norm 1 elements in $X_{p,w} \otimes X_{p,w}$ such that*
(a) *(z_n) is block diagonal, i.e., there exist increasing sequences of integers (m_j) and (p_j) such that $z_j = (Q_{m_{j+1}} - Q_{m_j}) \otimes (Q_{p_{j+1}} - Q_{p_j}) z_j$ for all j, where Q_i is the basis projection from $X_{p,w}$ onto the span of the first i basis elements,*
(b) *$(T z_n)$ is a block of the basis of $(\sum Y_n)_{p,2}$,*
(c) *there exists an increasing sequence of integers (M_n) such that*

$$\|(I - P_{M_n}) T z_{n-1}\| = 0$$

and

$$\|(I - P_{M_n})Tz_n\|_2 > \|T\|C\|z_n\|_2 \qquad \text{for all } n,$$

where P_i is the projection from $(\sum Y_n)_{p,2}$ onto $(\sum_{n=1}^{i} Y_n)_{p,2}$. Then

$$\left(\sum_{1}^{N} \|z_n\|_2^2\right)^{1/2} \le N^{1/p} \qquad \text{for all } N$$

Proof. We estimate the norms of $\sum z_n$ and $\sum Tz_n$. Clearly $\|\sum z_n\|_2 = (\sum \|z_n\|_2^2)^{1/2}$ and $\|\sum z_n\|_p \le (\sum \|z_n\|^p)^{1/p}$. Moreover, because the sequence (z_n) is block diagonal $(\sum \|z_n\|^p)^{1/p}$ dominates the mixed norms,

$$\|\sum_n z_n\|_R = (\sum_i \|\mathcal{R}_i \sum_n z_n\|_2^p)^{1/p} \text{ and } \|\sum_n z_n\|_C = (\sum_i \|\mathcal{C}_i \sum_n z_n\|_2^p)^{1/p}.$$

On the other hand if (r_n) is a sequence of Rademacher functions and $F \subset \mathbb{N}$ with $|F| = N$,

$$\|T\| \max\{N^{1/p}, (\sum_{n \in F} \|z_n\|^2)^{1/2}\} \ge \|T\|\|\sum_{n \in F} z_n\|$$

$$\ge \int \|\sum_{n \in F} r_n Tz_n\|_2$$

$$= (\sum_{n \in F} \|Tz_n\|_2^2)^{1/2}$$

$$\ge (\sum_{n \in F} \|(I - P_{M_n})Tz_n\|_2^2)^{1/2}$$

$$> C\|T\|(\sum_{n \in F} \|z_n\|^2)^{1/2}.$$

Therefore, $N^{1/p} \ge (\sum_{n \in F} \|z_n\|^2)^{1/2}$. $\square$

Let (w_n) be a sequence in $(0,1]$ which decreases to 0 and let $w_{n,k} = w_n$ for all $n, k \in \mathbb{N}$. Throughout the next few sections $(w_{n,k})$ will denote a doubly indexed sequence of this type. For the space $X_{p,2,(w_{n,k})}$ only some subsequences of the basis are again bases of $X_{p,2,(w_n')}$ for some (w_n') satisfying (*). The next definition contains a large enough class of such subsequences that we can restrict to this class for various gliding hump arguments.

DEFINITION 4.5. A subset S of $\mathbb{N} \times \mathbb{N}$ is said to be *rich* if there exists $M \subset \mathbb{N}$, M infinite, such that for every $m \in M$, $\{(m,k) \in S : k \in \mathbb{N}\}$ is infinite.

The rich sets are a fairly nice class from a combinatorial standpoint and we have already used them implicitly in the proof of Proposition 4.2. For example if S is a rich set and $A \subset S$ then either A or $S \setminus A$ must contain a rich set. Also if A contains a rich set, K, then there is a maximal rich set K' with $K \subset K' \subset A$.

Indeed $K' = \{(m, n) : |\{m\} \times \mathbb{N} \cap A| = \infty\}$. Another important point for us is that rich sets can be constructed by an induction procedure which imitates the usual method of showing that the rationals are countable.

In $X_p \otimes X_p$ each row and column is isomorphic to X_p, but $\| \cdot \|_2$ is affected by the choice of row or column. Previous results in this section show how the norm $\| \cdot \|_2$ in X_p (Lemma 4.1) or in $X_p \otimes X_p$ (Proposition 4.2) is "felt" by an operator at least on the basis vectors. In the next result we see that it also "felt" on each of the row and column spaces. This phenomenon is more subtle since it is caused by the other rows and columns.

PROPOSITION 4.6. *If (Y_n) is a sequence of subspaces of $L_p[0, 1]$ with $C-$ unconditional bases and $T : X_{p,(w_{n,k})} \otimes X_{p,(w_{n,k})} \to (\sum Y_n)_{p,2}$ is an isomorphism, then for every $\epsilon > 0$ there exists a rich subset K of $\mathbb{N} \times \mathbb{N}$ and for each $\kappa \in K$, there is an integer M_κ such that*

$$\|(I - P_{M_\kappa})Tz\|_2 \leq \|T\|Cw_\kappa(1 + \epsilon)\|z\|$$

for all $z \in [x_\kappa] \otimes X_{p,(w_{n,k})}$, where P_{M_κ} is the restriction operator from $(\sum_{n=1}^\infty Y_n)$ onto $(\sum_{n=1}^{M_\kappa} Y_n)$.

Proof. First we fix m, n and suppose that there is no such $M = M(m, n)$ for some ϵ and $[x_{m,n}] \otimes X_{p,(w_{n,k})}$. Then let (δ_i) be a sequence of positive numbers tending to 0. We may inductively choose a sequence of unit vectors $(z_{m,n,k})_{k=1}^\infty$ in $[x_{m,n}] \otimes X_{p,w}$ and an increasing sequence of positive integers (M_k) such that

$$\|(I - P_{M_k})Tz_{m,n,k-1}\| < \delta_k \qquad \text{for all } k$$

and

$$\|(I - P_{M_k})Tz_{m,n,k}\|_2 > \|T\|Cw_{m,n}.$$

By passing to a subsequence if necessary we may assume that there is a z_0 such that for each k, $z_{m,n,k} = z_0 + z'_{m,n,k}$, $(z'_{m,n,k})$ is a (perturbation of a) block of the basis of $[x_{m,n,k}] \otimes X_{p,(w_{n,k})}$, and $(Tz'_{m,n,k})$ is (a perturbation of) a block of the basis of $(\sum Y_m)_{p,2}$ which is disjoint from Tz_0. Observe that for sufficiently large k

$$\|(I - P_{M_k})Tz'_{m,n,k}\|_2 > (1 + \epsilon)\|T\|Cw_{m,n}\|z'_{m,n,k}\|.$$

Thus we may assume that $z_0 = 0$.

If for some m there are infinitely many n for which there is no $M_{m,n}$, then by a diagonalization argument we can find a sequence $(z_{m,n,k(n)})_{n \in N}$ and an increasing sequence of integers (M_n) such that

$$\|(I - P_{M_n})Tz_{m,n,k(n)}\| < \delta_n \qquad \text{for all } n$$
$$\|(I - P_{M_n})Tz_{m,n,k(n)}\|_2 > \|T\|Kw_{m,n}.$$

We may also assume that if $z_{m,n,k(n)} = x_{m,n} \otimes \zeta_n$ then (ζ_n) is a block of the basis of $X_{p,(w_{n,k})}$.

Because $w_{m,n}$ is the same for all n, Lemma 4.4 and a perturbation argument shows that this is impossible. Thus for any m there are only finitely many n for which the required integer $M_{m,n}$ does not exist, and K may be obtained by discarding these finitely many n for each m. □

We will now see that the estimate on the norm $\|\cdot\|_2$ actually reduces the tail to an ℓ_p sum.

COROLLARY 4.7. *Suppose that $(w_{n,k})$ is a sequence of positive numbers as above, T is an isomorphism from $X_{p,(w_{n,k})} \otimes X_{p,(w_{n,k})}$ into $(\sum Y_n)_{p,2}$ as in Proposition 4.6. Let S_M be the natural map from $(\sum Y_n)_{p,2}$ into $(\sum_{n=M+1}^{\infty} Y_n)_p$ and let P_M be the natural projection from $(\sum Y_n)_{p,2}$ onto $(\sum_{n=1}^{M} Y_n)_{p,2}$. Then there exists a rich set $K \subset \mathbb{N} \times \mathbb{N}$ and for each $\kappa \in K$ an integer M_κ such that $(S_{M_\kappa} + P_{M_\kappa})T$ is an isomorphism from $[e_\kappa] \otimes X_{p,w}$ into $(\sum_{n=1}^{M} Y_n)_{p,2} \oplus (\sum_{n=M_\kappa+1}^{\infty} Y_n)_p$, where $(e_{n,k})$ is the natural basis of $X_{p,(w_{n,k})}$.*

Proof. First we find a rich set K' as in Proposition 4.6 with $\epsilon < 1$. Because $w_n \downarrow 0$ there is a rich set $K \subset K'$ such that $4C\|T^{-1}\|\|T\|w_\kappa < 1$ for all $\kappa \in K$. From Proposition 4.6 we have that for $z \in [e_\kappa] \otimes X_{p,(w_{n,k})}$,

$$\|Tz\|_2^2 = \|(I - P_{M_\kappa})Tz\|_2^2 + \|P_{M_\kappa}Tz\|_2^2$$
$$\leq \|T\|^2 C^2 4 \sup w_\kappa^2 \|z\|_2^2 + \|P_{M_\kappa}Tz\|^2$$

(by Proposition 4.6 and $\|\cdot\|_2 \leq \|\cdot\|$ in $(\sum Y_n)_{p,2}$)

$$(4.7.1) \qquad \leq \|z\|_2^2/(\|T^{-1}\|^2 4) + \|P_{M_\kappa}Tz\|^2$$

since by the choice of K', $\|T\|^2 C^2 4 \sup w_\kappa^2 \leq \|T^{-1}\|^{-2}/4$. For any $z \in [e_\kappa] \otimes X_{p,(w_{n,k})}$,

$$\|T^{-1}\|^{-1}\|z\| \leq \|Tz\|$$
$$= \max\{\|Tz\|_2, (\sum \|(P_{n+1} - P_n)Tz\|_p^p)^{1/p}\}$$
$$\leq \max\{(\|T\|^2 C^2 4 \sup w_\kappa^2 \|z\|_2^2 + \|P_{M_\kappa}Tz\|^2)^{1/2},$$
$$\|P_{M_\kappa}Tz\| + (\sum_{n=M_\kappa+1}^{\infty} \|(P_{n+1} - P_n)Tz\|_p^p)^{1/p}\}.$$

If

$$\|P_{M_\kappa}Tz\| + (\sum_{n=M_\kappa+1}^{\infty} \|(P_{n+1} - P_n)Tz\|_p^p)^{1/p}\} \geq \|Tz\|,$$

we are done. If not,

$$(\|T\|^2 C^2 4 \sup w^2 \|z\|_2^2 + \|P_{M_\kappa}Tz\|^2)^{1/2} > \|Tz\| \geq \|T^{-1}\|^{-1}\|z\|$$

then by (4.7.1)

$$\|T^{-1}\|^{-2}\|z\|^2 \leq \|z\|_2^2/(\|T^{-1}\|^2 4) + \|P_{M_\kappa}Tz\|^2$$

and consequently

$$\|T^{-1}\|^{-2}\|z\|^2(3/4) \le \|P_{M_\kappa}Tz\|^2.$$

Thus

$$\|T\|\|T^{-1}\|\|P_{M_\kappa}Tz\|\sqrt{12}/3 \ge \|Tz\|$$

in this case.

It follows that $\|((S_{M_\kappa} + P_{M_\kappa})T)^{-1}\| \le \|T\|\|T^{-1}\|\sqrt{12}/3.$ $\square$

Remark 4.8. For a general isomorphism from $X_p \otimes X_p$ into $(\sum Y_n)_{p,2}$, we cannot hope to show that there is a bound $\infty > M \ge M_\kappa$. To see this notice that there is a simple isomorphic embedding of $X_{p,(w_n)} \otimes X_{p,(w_n)}$ into $(\sum X_{p,(w_n)})_{p,2,(w_n)} \oplus (\sum X_{p,(w_n)})_{p,2,(w_n)}$. Let (e_n) and (e'_n) be copies of the natural basis of $X_{p,(w_n)}$. Let $(d_{k,j})$ and $(d'_{k,j})$ be two copies of the natural basis of $(\sum X_{p,(w_n)})_{p,2,(w_n)}$. Define $T(e_n \otimes e'_m) = d_{n,m} \oplus d'_{m,n}$ and extend linearly. It is easy to see from Proposition 1.3 and the definition of the $(p,2)-$sum (or Corollary 2.9) that T is an isomorphism. Obviously no uniform bound on $(M_\kappa)_{\kappa \in K}$ can be found for this operator and any rich set K.

CHAPTER 5

SELECTION OF BASES IN $X_p \otimes X_p$

Because of the multi-index nature of the natural basis of $X_{p,(w_{n,k})} \otimes X_{p,(w_{n,k})}$ and the technical complexities of gliding hump type arguments with respect to a multi-index, we introduce a method of producing subsequences of the basis which still span a copy of $X_{p,(w_{n,k})} \otimes X_{p,(w_{n,k})}$ but which can be used without directly worrying about the nature of the underlying index set. We will use a fairly general setup in this section which may be applicable to other bases with complicated natural orderings.

DEFINITION 5.1. Let X be a Banach space with basis (x_i). A set $\mathcal{S}$ of infinite subsets of $\mathbb{N}$ is K-admissible for (x_i) if
(1) For each $n \in \mathbb{N}$ and $S \in \mathcal{S}$, $\{n, n+1, \dots\} \cap S \in \mathcal{S}$.
(2) $\mathbb{N} \in \mathcal{S}$.
(3) For each $S \in \mathcal{S}$, $(x_i)_{i \in S}$ is a basis for a subspace of X which is K-isomorphic to X.

DEFINITION 5.2. Suppose that X is a Banach space with basis (x_i), τ is a topology on X, and $\mathcal{S}$ is a admissible subset of $2^{\mathbb{N}}$ for (x_i). We will say that (x_i) has SP (selection property) with respect to τ and $\mathcal{S}$ if there is a winning strategy for the second player in the two player game described below.
(0) $S'_0 = \mathbb{N}$.
(1) On each turn n, $n = 1, 2, \dots$, the first player must define a multi-function F_n on X with range $F_n \subset \{0, 1, 2, \dots, N_n\}$, N_n finite, and $F_n^{-1}(j)$ is τ-open for each j and a set $S_n \in \mathcal{S}$ with $S_n \subset S'_{n-1}$ and $\{i_1, i_2, \dots, i_{n-1}\} \subset S_n$.
(2) On the turn n the second player must choose an integer $i_n \in S_n$, $i_n > i_{n-1}$, and a set $S'_n \in \mathcal{S}$ with $S'_n \subset S_n$ and $\{i_1, i_2, \dots, i_n\} \subset S'_n$.
Player 2 wins if (and only if) $\{x_{i_k} : k \in \mathbb{N}\}$ is in $\mathcal{S}$ and $F_n(x_{i_j})$ is constant for all $j \geq n$.

In the game the function F_n given by Player 1 defines a τ-open cover of X and Player 2 is forced to choose one of the sets, O, from the open cover and select all further elements for the subsequence from O. Usually it will not be necessary to define the multi-functions F_n on anything more than the τ-closure of the basis (x_i).

In X_p the most useful basis is indexed by $\mathbb{N} \times \mathbb{N}$. To treat such cases we will use an order like that used to enumerate the rationals. To this end let ϕ be a bijection from $\mathbb{N}$ onto $\mathbb{N} \times \mathbb{N}$ such that $\phi(j)^1 + \phi(j)^2 \leq \phi(i)^1 + \phi(i)^2$ if $j \leq i$. (Here we use superscripts to denote the coordinates of elements in $\mathbb{N} \times \mathbb{N}$, e.g., $(n, k)^1 = n$.) We order the basis of $X_{p,(w_{n,k})}$ as $(x_{\phi(i)})_{i=1}^{\infty}$. The admissible set $\mathcal{S}$ in this case is $\phi^{-1}(\{M \subset \mathbb{N} \otimes \mathbb{N} : \text{there is an infinite } N' \subset \mathbb{N} \text{ such that for each } n \in$

$N', (n, k) \in M$ for infinitely many k and $(n, k) \notin M$ for all $n \notin N'$ and $k \in \mathbb{N}\}$). Notice that these are the inverse images of the rich subsets of $\mathbb{N} \times \mathbb{N}$.

If M is an infinite subset of $\mathbb{N}$ and $i \in M$ then $o(i)$ will be the ordinal of the element in M as an ordered set with the inherited order. If there is some ambiguity about which set is under consideration, we will add the set as another parameter as in $o(i, M)$. The proof of the next lemma is presented in a very formal fashion. The idea of the proof is to follow a scheme like that used to enumerate the rationals. With this hint the reader may find it easier to generate his own proof than to work through this one.

LEMMA 5.3. *The standard basis of* $X_{p,(w_{n,k})}$, *where* $(w_{n,k})$ *satisfies* (*), *has the SP with respect to the weak topology and the class* $\mathcal{S}$ *defined above.*

Proof. Let $(x_{n,k})$ be the standard basis of $X_{p,(w_{n,k})}$. We will use induction to choose a subsequence $(x_{n,m})_{n \in \mathbb{N}, m \in M_n}$ with $\{(n, m) : n \in \mathbb{N}, m \in M_n\} \in \phi(\mathcal{S})$.

For notational convenience we will assume that $0 \in F_m^{-1}(0)$, for each m. If $A \in \phi(\mathcal{S})$, A^1 will denote $\{n : (n, k) \in A \text{ for some } k\}$, the projection of A into the first coordinate.

Consider F_1, S_1. We have assumed that $0 \in F_1^{-1}(0)$. Because w-$\lim_k x_{n,k} = 0$, for each $n \in \phi(S_1)^1$ there exists $M_{n,1} \subset (n \times \mathbb{N}) \cap \phi(S_1)$, $M_{n,1}$ infinite, such that $F_1(x_{n,k}) = 0$ for all n and all $k \in M_{n,1}$. Let $i_1 = \phi^{-1}((n_1, k))$ where $o(n_1, \phi(S_1)^1) = 1 = \phi(1)^1$ and $o(k, M_{n_1,1}) = 1$. Define $S_1' = \phi^{-1}(\cup_{n \in \phi(S_1)^1} M_{n,1})$.

Let Player 1 choose the multi-function F_2 and $S_2 \subset S_1'$, with $S_2 \in \mathcal{S}$. Because $F_2(0) = 0$ and w-$\lim_{k:(n,k) \in M_{n,1}} x_{n,k} = 0$, for each $n \in \phi(S_2)^1$ we can find an infinite subset $M_{n,2}$ of $M_{n,1} \cap \phi(S_2)$ such that $F_2(x_{n,k}) = 0$ and $i_1 < \phi^{-1}(n, k)$ for all $k \in M_{n,2}$. Let $i_2 = \phi^{-1}(n_2, k)$ where $o(n_2, \phi(S_2)^1) = \phi(2)^1$ and $o(k, M_{n_2,2}) = \phi(2)^2$ and let $S_2' = \phi^{-1}(\cup_{n \in \phi(S_2)^1} M_{n,2})$.

Suppose we have chosen $i_1 < i_2 < \cdots < i_{m-1}$, and $M_{n,m-1} = \phi(S_{m-1}') \cap (n \times \mathbb{N})$, for all $n \in \phi(S_m)^1$. Let Player 1 choose F_m and S_m. Because $F_m(0) = 0$ and w-$\lim_{k:(n,k) \in M_{n,m-1}} x_{n,k} = 0$, for each $n \in \phi(S_m)^1$ we can find an infinite subset $M_{n,m}$ of $M_{n,m-1} \cap \phi(S_m)$ such that $F_m(x_{n,k}) = 0$ and $i_{m-1} < \phi^{-1}(n, k)$ for all $k \in M_{n,m}$. Let $i_m = \phi^{-1}(n_m, k)$ where $o(n_m, \phi(S_m)^1) = \phi(m)^1$ and $o(k, M_{n_m,m}) = \phi(m)^2$ and let $S_m' = \phi^{-1}(\cup_{n \in \phi(S_m)^1} M_{n,m})$.

In this way the subsequence $(x_{\phi(i_j)})$ satisfies $F_m(x_{\phi(i_j)}) = 0$ for all $j \geq m$. Moreover since for any $j \geq m$, $o(\phi(i_m)^1, \phi(S_j)^1) = \phi(m)^1$ and $o(\phi(i_m)^2, \{\phi(i_j)^2 : \phi(i_j)^1 = \phi(m)^1\}) = \phi(i_m)^2 = o(\phi(m)^2, \mathbb{N})$, $\{i_j : j \in \mathbb{N}\}$ is in $\mathcal{S}$. $\square$

In the next result we pass to a tensor product. We will not define a class of admissible sets for the product index set but instead use the game in each factor to accomplish our goals. Below we use an unspecified tensor product of two Banach spaces. The only property that we require of the tensor product is that $\|x \otimes y\| \leq \|x\|\|y\|$ for all x, y.

PROPOSITION 5.4. *Suppose that* (x_i) *and* (y_j) *are shrinking unconditional bases with SP with respect to the weak topology and admissible sets* $\mathcal{S}(X)$ *and* $\mathcal{S}(Y)$, *respectively, and that* T *is a bounded operator from* $[x_i \otimes y_j : i, j \in \mathbb{N}]$ *into a space* Z *with normalized shrinking basis* (z_k). *Then given* $\epsilon > 0$ *there are* $I \in \mathcal{S}(X)$ *and* $J \in \mathcal{S}(Y)$ *and finite subsets* $N(i, j)$ *of* $\mathbb{N}$ *such that*

1) $\sum_{i \in I, j \in J} \|T(x_i \otimes y_j)|_{\mathbb{N} \setminus N(i,j)}\| < \epsilon$

2) $N(i,j) \cap N(i',j') = \emptyset$ if $i' \neq i$ and $j \neq j'$; $i = i'$, $j \neq j'$ and $\max(o(j), o(j')) > o(i)$; or $i \neq i'$, $j = j'$ and $\max(o(i), o(i')) \geq o(j)$.

Proof:. First let $\epsilon(i,j) > 0$ such that $\sum_{i,j} \epsilon(i,j) < \epsilon$, and $\epsilon(i,j)$ is decreasing, i.e., $\epsilon(i,j) \leq \epsilon(i',j')$ if $i \leq i', j \leq j'$. We will use two interweaving games to choose the sets I and J and use X and Y as subscripts on the associated functions to keep the notation straight. We begin the games with only trivial conditions: $F_{X,1}$ and $F_{Y,1}$ are constant functions and $S_{X,1} = S_{Y,1} = \mathbb{N}$. Let i_1, j_1 be the first elements chosen in each of the games and $S'_{X,1}, S'_{Y,1}$ be the resulting elements of $\mathcal{S}(X)$ and $\mathcal{S}(Y)$, respectively. Choose a finite subset $N(1,1)$ of $\mathbb{N}$ such that $\|T(x_{i_1} \otimes y_{j_1})|_{\mathbb{N} \setminus N(1,1)}\| < \epsilon(1,1)$.

Let $\eta_1 = \{1, 2, \ldots, \max N(1,1)\}$. Define $F_{Y,2}(x_{i_1}, y) = 0$ if $\sum_{k \in \eta_1} |z_k^*(Tx_{i_1} \otimes y)| < \epsilon(1,2)/2$ and $F_{Y,2}(y) = 1$ if $\sum_{k \in \eta_1} |z_k^*(Tx_{i_1} \otimes y)| > \epsilon(1,2)/3$. Let $S_{X,2} = S'_{X,1}$ and $S_{Y,2} = S'_{Y,1}$.

Player 2 must now select j_2 such that $F_{Y,2}(y_{j_2}) = 0$ because $(T(x_{i_1} \otimes y_j))$ converges to 0 weakly. Choose a finite subset $N(1,2)$ of $\mathbb{N} \setminus \eta_1$ such that $\|T(x_{i_1} \otimes y_{j_2})|_{\mathbb{N} \setminus N(1,2)}\| < \epsilon(1,2)$.

Let $\eta_2 = \{1, 2, \ldots, \max(N(1,1) \cup N(1,2))\}$,

$$F_{X,2}(x) = 0 \text{ if } \max_{s=1,2} \sum_{k \in \eta_2} |z_k^*(T(x \otimes y_{j_s}))| < \epsilon(2,2)/2$$

and

$$F_{X,2}(x) = 1 \text{ if } \max_{s=1,2} \sum_{k \in \eta_2} |z_k^*(T(x \otimes y_{j_s}))| > \epsilon(2,2)/3.$$

Let $S_{X,2} = S'_{X,1}$.

Player 2 in the X game must choose i_2 such that $F_{X,2}(x_{i_2}) = 0$. For $k = 1, 2$ choose a finite set $N(2,k) \subset \mathbb{N} \setminus \eta_2$, such that $\|T(x_{i_2} \otimes y_{j_k})|_{\mathbb{N} \setminus N(2,k)}\| < \epsilon(2,k)$.

Let $\eta_3 = \{1, 2, \ldots, \max \cup_{s \leq 2, t \leq 2} N(s,t)\}$,

$$F_{Y,3}(y) = 0 \text{ if } \max_{n'=1,2} \sum_{k \in \eta_3} |z_k^*(T(x_{i_{n'}} \otimes y))| < \epsilon(2,3)/2$$

and

$$F_{Y,3}(y) = 1 \text{ if } \max_{n'=1,2} \sum_{k \in \eta_3} |z_k^*(T(x_{i_{n'}} \otimes y))| > \epsilon(2,3)/3.$$

Let $S_{Y,3} = S'_{Y,2}$.

Player 2 in the Y game must choose j_3 such that $F_{Y,3}(y_{j_3}) = 0$. For $k = 1, 2$ choose a finite set $N(k,3) \subset \mathbb{N} \setminus \eta_3$ such that $\|T(x_{i_k} \otimes y_{j_3})|_{\mathbb{N} \setminus N(k,3)}\| < \epsilon(k,3)$.

This completes the first few steps of the induction. Assume that $i_1, \ldots, i_r$ and $j_1, \ldots j_{r+1}$ are known and the corresponding sets $N(n,m), n = 1, 2, \ldots r, m = 1, 2, \ldots, r+1$ have been chosen.

We continue with the $r + 1$ turn of the X game. Let

$$\eta_{2r} = \{1, 2, \ldots, \max \cup_{s \leq r, t \leq r+1} N(s,t)\}$$

and

$$F_{X,r+1}(x) = 0 \qquad \text{if } \max_{s \leq r+1} \sum_{k \in \eta_{2r}} |z_k^*(T(x \otimes y_{j_s}))| < \epsilon(r+1, r+1)/2$$

and

$$F_{X,r+1}(x) = 1 \text{ if } \max_{s \leq r+1} \sum_{k \in \eta_{2r}} |z_k^*(T(x \otimes y_{j_s}))| > \epsilon(r+1, r+1)/3.$$

Let $S_{X,r+1} = S'_{X,r}$.

Player 2 in the X game must choose i_{r+1} such that $F_{X,r+1}(x_{i_{r+1}}) = 0$. For $k = 1, 2, \ldots, r+1$ choose a finite set $N(r+1, k) \subset \mathbb{N} \setminus \eta_{2r}$ such that $\|T(x_{i_2} \otimes y_{j_k})|_{\mathbb{N} \setminus N(r+1,k)}\| < \epsilon(r+1, k)$.

Let $\eta_{2r+1} = \{1, 2, \ldots, \max \cup_{s \leq r+1, t \leq r+1} N(s,t)\}$ and

$$F_{Y,r+2}(y) = 0 \text{ if } \max_{n' \leq r+1} \sum_{k \in \eta_{2r+1}} |z_k^*(T(x_{i_{n'}} \otimes y))| < \epsilon(r+1, r+2)/2$$

and

$$F_{Y,r+2}(y) = 1 \text{ if } \max_{n' \leq r+1} \sum_{k \in \eta_{2r+1}} |z_k^*(T(x_{i_{n'}} \otimes y))| > \epsilon(r+1, r+2)/3$$

Let $S_{Y,r+2} = S'_{Y,r+1}$.

Player 2 in the Y game must choose j_{r+2} such that $F_{Y,r+2}(y_{j_{r+2}}) = 0$. For $k = 1, 2, \ldots, r+1$ choose a finite set $N(k, r+2) \subset \mathbb{N} \setminus \eta_{2r+1}$ such that

$$\|T(x_{i_k} \otimes y_{j_{r+2}})|_{\mathbb{N} \setminus N(k,r+2)}\| < \epsilon(k, r+2).$$

It is easy to see that we have now completed the next step of the induction and the result follows. $\square$

CHAPTER 6

$X_p \otimes X_p$-**PRESERVING OPERATORS ON** $X_p \otimes X_p$

The purpose of the section is to prove a criterion which guarantees that an operator on $X_p \otimes X_p$ preserves isomorphically a copy of the space. Similar results exist for spaces with an unconditional basis which have many subsequences equivalent to the original basis. The tensor product makes the combinatorics more difficult here. In Proposition 5.4 we were unable to get completely disjoint blocks. In order to use the SP for the basis of each factor of $X_p \otimes X_p$ to get really disjoint blocks it is necessary to have some quantitative information. The next two lemmas give us estimates of how many vectors we will need in order to insure that we can get reasonable disjointness for a step of a gliding hump argument.

In the next lemma we make use of the concept of a lower ℓ_r-estimate [LTII, Definition 1.f.4]. We say that a basic sequence (z_n) has a lower ℓ_r-estimate with constant C if (and only if)

$$\| \sum a_n z_n \| \geq C (\sum |a_n|^r)^{1/r}$$

for every sequence of real numbers (a_n).

LEMMA 6.1. *If T is an operator from $X_{p,(w_n)}$ into a space Z with a basis (z_n) and normalized biorthogonal functionals (z_n^*) such that for some $r \geq 2$, (z_n) satisfies a lower ℓ_r estimate with constant C, then for any $k \in \mathbb{N}$ and ϵ, $\|T\| > \epsilon > 0$,*

$$\sum_{n \in F} w_n^{2p/(p-2)} \leq \|T\|^r 2k^{r+1}/(\epsilon^r C^r \min\{w_i^r : 1 \leq i \leq k\}),$$

where (x_n) is the standard basis of $X_{p,(w_n)}$ and

$$F = \{n : |z_i^*(Tx_n)| > \frac{\epsilon w_n}{k} \text{ or } |z_n^*(Tx_i)| > \frac{\epsilon w_i}{k} \text{ for some } i \leq k\}.$$

Proof. For $1 \leq i \leq k$ let $F_i = \{n : |z_i^*(Tx_n)| > \frac{\epsilon w_n}{k}\}$ and let $G_i = \{n : |z_n^*(Tx_i)| > \frac{\epsilon w_i}{k}\}$. Then for $n \in F$, $n \in F_i$ or G_i for at least one i. For $n \in F_i$ let $a_n = w_n^{(p+2)/(p-2)}$. Then

$$\| \sum_{n \in F_i} a_n x_n \| = \max\{(\sum_{n \in F_i} a_n^2 w_n^2)^{1/2}, (\sum_{n \in F_i} a_n^p)^{1/p}\}$$

$$= \max\{(\sum_{n \in F_i} w_n^{4p/(p-2)})^{1/2}, (\sum_{n \in F_i} w_n^{p(p+2)/(p-2)})^{1/p}\}$$

$$\leq \max\{(\sum_{n \in F_i} w_n^{2p/(p-2)})^{1/2}, (\sum_{n \in F_i} w_n^{2p/(p-2)})^{1/p}\}.$$

Let $W_i = \sum_{n \in F_i} w_n^{2p/(p-2)}$. For each i and choice of signs σ_n,

$$|z_i^*(T \sum_{n \in F_i} \sigma_n a_n x_n)| \leq \|T\| \max\{W_i^{1/2}, W_i^{1/p}\}.$$

Thus

$$\epsilon W_i/k = \sum_{n \in F_i} \epsilon w_n^{2p/(p-2)}/k = \sum_{n \in F_i} \epsilon w_n |a_n|/k \leq \|T\| \max\{W_i^{1/2}, W_i^{1/p}\}.$$

If $W_i > 1$ then $W_i^{1/2} > W_i^{1/p}$ and hence $W_i \leq \|T\|^2 k^2/\epsilon^2$; if $W_i \leq 1$,

$$W_i \leq \|T\|^{p/(p-1)} k^{p/(p-1)}/\epsilon^{p/(p-1)} \leq \|T\|^2 k^2/\epsilon^2$$

also.

For each $i \leq k$,

$$\|Tx_i\| \geq C(\sum_{n \in G_i} |z_n^*(Tx_i)|^r)^{1/r} \geq C\epsilon w_i k^{-1} |G_i|^{1/r}$$

$$\geq C\epsilon w_i k^{-1} (\sum_{n \in G_i} w_n^{2p/(p-2)})^{1/r}.$$

Thus $\sum_{n \in G_i} w_n^{2p/(p-2)} \leq k^r \|T\|^r/(\epsilon^r C^r w_i^r)$. Let $W_i' = \sum_{n \in G_i} w_n^{2p/(p-2)}$.
Then

$$\sum_{n \in F} w_n^{2p/(p-2)} \leq \sum_{i=1}^k W_i + W_i'$$

$$\leq \sum_{i=1}^k (\frac{\|T\|^2 k^2}{\epsilon^2} + \frac{\|T\|^r k^r}{\epsilon^r C^r w_i^r})$$

$$\leq \frac{\|T\|^r 2k^{r+1}}{\epsilon^r C^r \min\{w_i^r : 1 \leq i \leq k\}}.$$

$\square$

Notice that if a finite sequence of positive numbers (ϵ_i) is specified in advance and some control on $\min\{w_i^r : 1 \leq i \leq k\}$ is given, it is possible to predict the number of elements required to produce an approximately blocked image. To be more precise we have the following.

LEMMA 6.2. *Let (ϵ_i) be a sequence of positive numbers, $C, D, w_0 \in \mathbb{R}^+$ and $K \in \mathbb{N}$. There exists an integer N_0, such that if T is an operator from $X_{p,(w_n)_{n=1}^{N_0}}$, with standard basis (x_n), into a space Z with a basis (z_n) and biorthogonal functionals (z_n^*) such that for some $r \geq 2$, (z_n) satisfies a lower ℓ_r estimate with constant C, $w_n \geq w_0$ for all $n \leq N$, and $\|T\| \leq D$, then there exist $\{n_j : 1 \leq j \leq K\}$ such that for $1 \leq i < j \leq K$,*

$$|z_{n_i}^*(Tx_{n_j})| < \epsilon_j w_{n_j}/(j-1)$$

and

$$|z^*_{n_j}(Tx_{n_i})| < \epsilon_i w_{n_i}/(j-1).$$

Proof. We use induction on K. Notice that if $K = 1$, the requirements are vacuous, i.e., $N_0 = 1$ works. If true for K, let $N(K)$ denote the required integer. We need a sufficiently large $N(K+1)$ to apply Lemma 6.1 with $\epsilon = \epsilon_{K+1}$ (We may assume that $\epsilon_{K+1} = \min\{\epsilon_j : 1 \le j \le K+1\}$.), $k = K$, $z_i = z_{n_i}$, and the basis of $X_{p,(w_n)}$ reordered so that x_{n_i} is the ith basis vector and the remaining basis vectors follow these. Then, because $\sum_{n=K+1}^{N(K+1)} w_n^{2p/(p-2)} \ge \sum_{n=K+1}^{N(K+1)} w_0^{2p/(p-2)}$, if $(N(K+1) - (K+1))w_0^{2p/(p-2)} > D^r 2 K^{r+1}/(\epsilon_{K+1} w_0 C)^r$, the inductive hypothesis gives us $\{n_i : 1 \le i \le k\}$ and Lemma 6.2 will produced $n_{k+1} \in \{K+1, \ldots, N(K+1)\}$. $\square$

With these lemmas we can now show that operators on $X_p \otimes X_p$ with a significant diagonal are isomorphisms on a subspace isomorphic to the whole space. The proof makes use of the basic technique of [CL]. The problem here is to overcome the technical difficulties in getting a large block basic sequence. Below the notation is rather cumbersome so we have omitted parentheses where there should be no cause for confusion. Thus $T(z \otimes w)$ may be written as $Tz \otimes w$ and $(x^* \otimes y^*)(Tu)$ as $x^* \otimes y^*(Tu)$.

PROPOSITION 6.3. *Suppose that T is a bounded operator on $X_{p,(w_{n,k})} \otimes X_{p,(w_{n,k})}$ such that there exists $\epsilon > 0$ with $|(x^*_{n,k} \otimes y^*_{m,j})(Tx_{n,k} \otimes y_{m,j})| \ge \epsilon$ for all n, k, m, j. Then there are rich subsets K, J of $\mathbb{N} \times \mathbb{N}$ such that*

$$T|_{[x_{n,k} \otimes y_{m,j} : (n,k) \in K, (m,j) \in J]}$$

is an isomorphism. Moreover, $T([x_{n,k} \otimes y_{m,j} : (n,k) \in K, (m,j) \in J])$ is complemented.

Proof. First observe that if J, K are rich subsets of $\mathbb{N} \times \mathbb{N}$ and

$$Z = [x_{n,k} \otimes y_{m,j} : (n,k) \in K, (m,j) \in J],$$

then we can restrict our attention to Z. Indeed, because the basis of $X_{p,(w_{n,k})} \otimes X_{p,(w_{n,k})}$ is unconditional, we can compose $T|_Z$ with the basis projection P onto Z and a diagonal operator D on Z such that $D(x_{n,k} \otimes y_{m,j}) = ((x^*_{n,k} \otimes y^*_{m,j})(Tx_{n,k} \otimes y_{m,j}))^{-1} x_{n,k} \otimes y_{m,j}$ for all n, k, m, j. The result will be obtained by showing that K, J can be chosen so that the resulting composition is a perturbation of the identity on Z. As in [CL] this will imply that $T(Z)$ is complemented since for any $z \in Z$, $(TDP)Tz = T(DPTz) \approx Tz$.

By Proposition 5.4 for any $\delta > 0$ we may find rich sets K', J' and finite subsets of $\mathbb{N}^4$, $N_{k,j}, k \in K', j \in J'$ such that
(1) $\sum_{k \in K', j \in J'} \|T(x_k \otimes y_j)|_{\mathbb{N}^4 \setminus N_{k,j}}\| < \delta$
(2) $N(k,j) \cap N(k',j') = \emptyset$ if $k' \ne k$ and $j \ne j'$; if $k = k'$ and $\max(o(j), o(j')) > o(k)$; or if $j = j'$ and $\max(o(k), o(k')) \ge o(j)$.

If δ is sufficiently small, it follows from standard arguments that if $\eta_n = \{j : o(j) \leq o(n)\}$ and $\eta'_m = \{k : o(k) < o(m)\}$ for all $n \in K', m \in J'$, then $T(Z)$ is isomorphic to the UFDD

$$\sum_{n \in K'} [T(x_n \otimes y_j)|_{\cup \{N(n,j): j \in \eta_n\}} : j \in \eta_n] \oplus \sum_{m \in J'} [T(x_k \otimes y_m)|_{\cup \{N(k,m): k \in \eta'_m\}} : k \in \eta'_m].$$

In particular, if $\delta < \epsilon$, then $(i,j) \in N(i,j)$ for all i,j. We need to refine the index sets J', K' further to get an actual (perturbation of a) basis rather than a UFDD. To save the notational burden of carrying a set of error terms through the computation we will assume that $\|T(x_k \otimes y_j)|_{\mathbb{N}^4 \setminus N_{k,j}}\| = 0$ for all $k \in K', j \in J'$.

Let (ϵ_i) be a sequence of positive numbers such that $\sum \epsilon_i < \epsilon'/4^p$, where $\epsilon' < \min(1, \epsilon)$. We will use the estimate in Lemma 6.1 repeatedly in the proof, but its exact nature is not used. The important thing is the dependence only on $r, C, \epsilon, k, \min\{w_i : i \leq k\}$, and $\|T\|$. In this proof $r = p$, $C = 1$ and $\|T\|$ are fixed, so we will denote the estimate by $H(\epsilon, w, k)$, where $w = \min\{w_i : i \leq k\}$.

We will essentially follow the two game argument of Proposition 5.4 but this time make more careful choices. We will again make use of the admissible class $\mathcal{S}$ for $X_{p,(w_{n,k})}$. As before the selection of i_1, j_1 is not controlled. The choice j_2 is more critical since we must make the contribution to the support of $Tx_{\phi(i)} \otimes y_{\phi(j_1)}$ small for all i in some set in the class $\mathcal{S}(X)$.

For each $j \in S'_{Y,1}$, $j > j_1$, let

$$K_j = \{i : i \in S'_{X,2}, |(x^*_{\phi(i)} \otimes y^*_{\phi(j_1)})(Tx_{\phi(i)} \otimes y_{\phi(j)})| < \epsilon_1 w_{\phi(j)},$$
$$\text{and } |(x^*_{\phi(i)} \otimes y^*_{\phi(j)})(Tx_{\phi(i)} \otimes y_{\phi(j_1)})| < \epsilon_1 w_{\phi(j_1)}\}.$$

We need to show that $\{j : K_j \supset K'_j \ni i_1 \in K'_j \text{ and } K'_j \in \mathcal{S}(X)\} \cup \{j_1\}$ contains a set $S_{Y,2} \in \mathcal{S}(Y)$, with $j_1 \in S_{Y,2}$ for which $\phi(S_{Y,2}) \cap (\{\phi(j_1)^1\} \times \mathbb{N})$ is infinite. We will first show that for almost all j there exists $K'_j \subset K_j$ such that $K'_j \in \mathcal{S}(X)$. For each i, by applying Lemma 6.1 to the operator $T_i : [x_{\phi(i)} \otimes y_{\phi(j)} : j \in \mathbb{N}] \to [x_{\phi(i)} \otimes y_{\phi(j)} : j \in \{j_1\} + (\mathbb{N} \setminus \{j_1\})]$ defined by

$$T_i(z) = x^*_{\phi(i)} \otimes y^*_{\phi(j_1)}(Tz)x_{\phi(i)} \otimes y_{\phi(j_1)} + \sum_{j \neq j_1} x^*_{\phi(i)} \otimes y^*_{\phi(j)}(Tz)x_{\phi(i)} \otimes y_{\phi(j)},$$

we have that $\sum_{j: i \notin K_j} w^{2p/(p-2)}_{\phi(j)} < H(\epsilon_1, \{w_{\phi(j_1)}\}, 1)$. If for some j, K_j contains no subset which is in $\mathcal{S}(X)$, then $\phi(K_j) \cap (\{m\} \times \mathbb{N})$ is finite for all but finitely many m. Lemma 6.1 implies that $\sum w^{2p/(p-2)}_{\phi(j)}$ over such j must be no more than $H(\epsilon_1, \{w_{\phi(j_1)}\}, 1)$. Indeed, if the sum exceeded this bound then there would be a finite set F of such j such that $\sum_{j \in F} w^{2p/(p-2)}_{\phi(j)} > H(\epsilon_1, \{w_{\phi(j_1)}\}, 1)$. Then since no subset of K_j is in $\mathcal{S}(X)$, there would be infinitely many m for which $\phi(K_j) \cap (\{m\} \times \mathbb{N})$ is finite for all of those $j \in F$. Thus for any $i \in S_{X,2} \setminus \cup_{j \in F} K_j$, T_i would violate the conclusion of Lemma 6.1. In particular for those j such that $\phi(j) \in (\{\phi(j_1)^1\} \times \mathbb{N})$, we have that $w_{\phi(j)} = w_{\phi(j_1)}$ and thus only finitely many of these K_j will fail to contain a subset K'_j as above. Similarly, if $F \subset \{j : K_j \not\supset K'_j \ni i_1 \in K'_j, K'_j \in \mathcal{S}(X)$ but $\exists K \subset K_j, K \in S(Y)\}$ and $\sum_{j \in F} w_{\phi(j)}^{2p/(p-2)} > H(\epsilon_1, \{w_{\phi(j_1)}\}, 1)$ then $\phi(K_j) \cap (\{\phi(i_1)^1\} \times \mathbb{N})$ is finite for all of those $j \in F$. Thus for any $i \in S'_{X,2} \setminus \cup_{j \in F} K_j$ with $\phi(i)^1 = \phi(i_1)^1$, T_i would violate the conclusion of Lemma 6.1. It follows that the required set $S_{Y,2}$ exists.

Player 2 selects j_2 from $S_{Y,2}$ and selects $S'_{Y,2} \subset S_{Y,2}$ which is in $\mathcal{S}(Y)$ and contains $\{j_1, j_2\}$. By our choice of $S_{Y,2}$, K_{j_2} contains a set K'_{j_2} such that $i_1 \in K'_{j_2} \in \mathcal{S}(X)$. Observe that this implies that

$$|(x^*_{\phi(i_1)} \otimes y^*_{\phi(j_1)})(T x_{\phi(i_1)} \otimes y_{\phi(j_2)})| < \epsilon_1 w_{\phi(j_2)},$$

and

$$|(x^*_{\phi(i_1)} \otimes y^*_{\phi(j_2)})(T x_{\phi(i_1)} \otimes y_{\phi(j_1)})| < \epsilon_1 w_{\phi(j_1)}.$$

The set $S = K'_{j_2}$ is our candidate for $S_{X,2}$ but we must refine it a little.

For each $i \in S'_{X,1}$ let

$$N_i = \{j : j \in S'_{Y,2}, |(x^*_{\phi(i_1)} \otimes y^*_{\phi(j)})(T x_{\phi(i)} \otimes y_{\phi(j)})| < \epsilon_1 w_{\phi(i)}$$
$$\text{and } |(x^*_{\phi(i)} \otimes y^*_{\phi(j)})(T x_{\phi(i_1)} \otimes y_{\phi(j)})| < \epsilon_1 w_{\phi(i_1)}\}.$$

We need to show that $\{i : N_i \supset N'_i \ni j_1, j_2 \in N'_i, N'_i \in \mathcal{S}(Y)\} \cup \{i_1\}$ contains a set $S_{X,2} \in \mathcal{S}(X)$, $i_1 \in S_{X,2}$, for which $\phi(S_{X,2}) \cap (\{\phi(i_1)^1\} \times \mathbb{N})$ is infinite. First observe that for i sufficiently large $j_1, j_2 \in N_i$. Next we will show that for almost all i there exists $N'_i \subset N_i$ such that $N'_i \Subset \mathcal{S}(Y)$. By applying Lemma 6.1, for each j, to $T_j : [x_{\phi(i)} \otimes y_{\phi(j)} : j \in \mathbb{N}] \to [x_{\phi(i_1)} \otimes y_{\phi(j)} : i \in \{i_1\} + (\mathbb{N} \setminus \{i_1\})]$ defined by

$$T_i(z) = x^*_{\phi(i_1)} \otimes y^*_{\phi(j)}(Tz)x_{\phi(i_1)} \otimes y_{\phi(j)} + \sum_{i \neq i_1} x^*_{\phi(i)} \otimes y^*_{\phi(j)}(Tz)x_{\phi(i)} \otimes y_{\phi(j)},$$

we have that $\sum_{i:j \notin N_i} w_{\phi(i)}^{2p/(p-2)} < H(\epsilon_1, \{w_{\phi(i_1)}\}, 1)$. If for some i, N_i contains no subset which is in $\mathcal{S}(Y)$, then $\phi(N_i) \cap (\{m\} \times \mathbb{N})$ is finite for all but finitely many m. Lemma 6.1 implies that $\sum w_{\phi(i)}^{2p/(p-2)}$ over such i must be no more than $H(\epsilon_1, \{w_{\phi(i_1)}\}, 1)$. Indeed if the sum exceeded this bound then there would be a finite set F of such i such that $\sum_{i \in F} w_{\phi(i)}^{2p/(p-2)} > H(\epsilon_1, \{w_{\phi(i_1)}\}, 1)$. Then there would be infinitely many m for which $\phi(N_i) \cap (\{m\} \times \mathbb{N})$ is finite for all of those $i \in F$, thus for any $j \in S'_{Y,2} \setminus \cup_{i \in F} N_i$, T_j would violate the conclusion of Lemma 6.1. In particular for those i such that $\phi(i) \in (\{\phi(i_1)^1\} \times \mathbb{N})$, $w_{\phi(i)} = w_{\phi(i_1)}$ and thus only finitely many of these N_i will fail to contain a subset N'_i as above. Similarly, if $F \subset \{i : N_i \not\supset N'_i \ni j_2 \in N'_i, \{j_1\} \cup N'_i \in \mathcal{S}(Y)$ but $\exists N \subset N_i, N \in S(Y)\}$ and $\sum_{i \in F} w_{\phi(i)}^{2p/(p-2)} > 2H(\epsilon_1, \{w_{\phi(i_1)}\}, 1)$, then $\phi(N_i) \cap (\{\phi(j)^1\} \times \mathbb{N})$ is finite for $j = j_1$ or j_2 for all of those $i \in F$, thus for at least one of $j' = j_1, j_2$ there exists $F' \subset F$ such that $\sum_{i \in F'} w_{\phi(i)}^{2p/(p-2)} > H(\epsilon_1, \{w_{\phi(i_1)}\}, 1)$. and $j \in S'_{Y,2} \setminus \cup_{i \in F} N_i$ with $\phi(j)^1 = \phi(j')^1$. This would violate the conclusion of Lemma 6.1 for T_j. It follows that the required set $S_{X,2}$ exists.

Player 2 selects i_2 from $S_{X,2}$ and selects $S'_{X,2} \subset S_{X,2}$ which is in $\mathcal{S}(X)$ and contains $\{i_1, i_2\}$. By our choice of $S_{X,2}$, N_{i_2} contains N'_{i_2} such that $\{j_1, j_2\} \subset N'_{i_2} \in \mathcal{S}(Y)$. Note that for all $j \in N'_{i,2}$, and $s \neq t \in \{1, 2\}$,

$$|(x^*_{\phi(i_s)} \otimes y^*_{\phi(j)})(Tx_{\phi(i_t)} \otimes y_{\phi(j)})| < \epsilon_{(s \vee t)-1} w_{\phi(i_t)}.$$

The set $S = N'_{i_2}$ is our candidate for $S_{Y,3}$ however we must now refine it further as we did above to produce $S_{X,2}$. This completes the initial phase of the induction.

Suppose that we have played r turns of each game to get $I = \{i_1, \ldots, i_r\}$, $J = \{j_1, \ldots, j_r\}$, $J \subset S'_{Y,r} \in \mathcal{S}(Y)$ and $I \subset S'_{X,r} \in \mathcal{S}(X)$.

For each $j \in S'_{Y,r} \setminus J$ let

$$K_j = \{i : i \in S'_{X,r}, |(x^*_{\phi(i)} \otimes y^*_{\phi(j')})(Tx_{\phi(i)} \otimes y_{\phi(j)})| < \epsilon_r w_{\phi(j)}/r$$
$$\text{and } |(x^*_{\phi(i)} \otimes y^*_{\phi(j)})(Tx_{\phi(i)} \otimes y_{\phi(j')})| < \epsilon_r w_{\phi(j')}/r \text{ for all } j' \in J\}.$$

We need to show that $\{j : K_j \supset K'_j \ni I \subset K'_j \text{ and } K'_j \in \mathcal{S}(X)\} \cup J$ contains a set $S_{Y,r+1} \in \mathcal{S}(Y)$, $J \subset S_{Y,r+1}$ for which $\phi(S_{Y,r+1}) \cap (\{\phi(j)^1\} \times \mathbb{N})$ is infinite for all $j \in J$. We will show that for almost all j there exists $K'_j \subset K_j$ such that $K'_j \in \mathcal{S}(X)$. For each i let the operator $T_i : [x_{\phi(i)} \otimes y_{\phi(j)} : j \in \mathbb{N}] \to [x_{\phi(i)} \otimes y_{\phi(j)} : j \in J + (\mathbb{N} \setminus J)]$, where the notation $J + (\mathbb{N} \setminus J)$ means that the elements in J precede the others, be defined by

$$T_i(z) = \sum_{j \in J} x^*_{\phi(i)} \otimes y^*_{\phi(j)}(Tz)x_{\phi(i)} \otimes y_{\phi(j)} + \sum_{j \notin J} x^*_{\phi(i)} \otimes y^*_{\phi(j)}(Tz)x_{\phi(i)} \otimes y_{\phi(j)}.$$

By Lemma 6.1 for each i, $\sum_{j:i \notin K_j} w^{2p/(p-2)}_{\phi(j)} < H(\epsilon_r, \min\{w_{\phi(j')} : j' \in J\}, r)$ If for some j, K_j contains no subset which is in $\mathcal{S}(X)$, then $\phi(K_j) \cap (\{m\} \times \mathbb{N})$ is finite for all but finitely many m. Lemma 6.1 implies that $\sum w^{2p/(p-2)}_{\phi(j)}$ over such j must be no more than $H(\epsilon_r, \min\{w_{\phi(j')} : j' \in J\}, r)$. Indeed if the sum exceeded this bound then there would be a finite set F of such j such that $\sum_{j \in F} w^{2p/(p-2)}_{\phi(j)} > H(\epsilon_r, \min\{w_{\phi(j')} : j' \in J\}, r)$. Then there would be infinitely many m for which $\phi(K_j) \cap (\{m\} \times \mathbb{N})$ is finite for all of those $j \in F$. Thus any $i \in S'_{X,r} \setminus \cup_{j \in F} K_j$ would violate the conclusion of Lemma 6.1. In particular for those j such that $\phi(j) \in (\{\phi(j')^1\} \times \mathbb{N})$, for some $j' \in J$, $w_{\phi(j)} = w_{\phi(j')}$ and thus only finitely many of these K_j will fail to contain a subset K'_j as above. Similarly, if $F \subset \{j : K_j \not\supset K'_j \ni I \subset K'_j, K'_j \in \mathcal{S}(X) \text{ but } \exists K \subset K_j, K \in \mathcal{S}(X)\}$ and $\sum_{j \in F} w^{2p/(p-2)}_{\phi(j)} > rH(\epsilon_r, \min\{w_{\phi(j')} : j' \in J\}, r)$ then for each $i \in I$, let $F_i = \{j \in F : \phi(K_j) \cap (\{\phi(i)^1\} \times \mathbb{N}) \text{ is finite}\}$. For at least one $i' \in I$, $\sum_{j \in F_i} w^{2p/(p-2)}_{\phi(j)} > H(\epsilon_r, \min\{w_{\phi(j')} : j' \in J\}, r)$ thus any $i \in S'_{X,2} \setminus \cup_{j \in F_i} K_j$ with $\phi(i)^1 = \phi(i')^1$ would violate the conclusion of Lemma 6.1. It follows that the required set $S_{Y,r+1}$ exists.

Player r+1 selects j_{r+1} from $S_{Y,r+1}$ and selects $S'_{Y,r+1} \subset S_{Y,r+1}$ which is in $\mathcal{S}(Y)$ and contains $J \cup \{j_{r+1}\}$. By our choice of $S_{Y,r+1}$, $K_{j_{r+1}}$ contains $K'_{j_{r+1}}$ such that $I \subset K'_{j_{r+1}} \in \mathcal{S}(X)$. Note that for all $i \in K'_{j,r+1}$, and $s \neq t \in \{1, 2, \ldots, r+1\}$,

$$|(x^*_{\phi(i)} \otimes y^*_{\phi(j_s)})(Tx_{\phi(i)} \otimes y_{\phi(j_t)})| < \epsilon_{((s \vee t))-1} w_{\phi(j_t)}.$$

The set $S = K'_{j_{r+1}}$ is our candidate for $S_{X,r+1}$ but we must again refine it a little.

For each $i \in S \setminus I$ let

$$N_i = \{k : k \in S'_{Y,r+1}, |x^*_{\phi(i')} \otimes y^*_{\phi(k)}(Tx_{\phi(i)} \otimes y_{\phi(k)})| < \epsilon_r w_{\phi(i)}/r$$
$$\text{and } |(x^*_{\phi(i)} \otimes y^*_{\phi(j)})(Tx_{\phi(i')} \otimes y_{\phi(j)})| < \epsilon_r w_{\phi(i')}/r \text{ for all } i' \in I\}$$

We need to show that $\{i : N_i \supset N'_i \ni J \subset N'_i, N'_i \in \mathcal{S}(Y)\} \cup I$ contains a set $S_{X,r+1} \in \mathcal{S}(X)$, $I \subset S_{X,r+1}$, for which $\phi(S_{X,r+1}) \cap (\{\phi(i)^1\} \times \mathbb{N})$ is infinite for all $i \in I$. First observe that for i sufficiently large $J \subset N_i$. Next we will show that for almost all i there exists $N'_i \subset N_i$ such that $N'_i \in \mathcal{S}(Y)$. By applying Lemma 6.1 for each j, as above, $\sum_{i:j \notin N_i} w_{\phi(i)}^{2p/(p-2)} < H(\epsilon_r, \min\{w_{\phi(i')} : i' \in I\}, r)$. If for some i, N_i contains no subset which is in $\mathcal{S}(Y)$, then $\phi(N_i) \cap (\{m\} \times \mathbb{N})$ is finite for all but finitely many m. Lemma 6.1 implies that $\sum w_{\phi(i)}^{2p/(p-2)}$ over such i must be no more than $H(\epsilon_r, \min\{w_{\phi(i')} : i' \in I\}, r)$. Indeed if the sum exceeded this bound then there would be a finite set F of such i such that $\sum_{i \in F} w_{\phi(i)}^{2p/(p-2)} > H(\epsilon_r, \min\{w_{\phi(i')} : i' \in I\}, r)$. Then there would be infinitely many m for which $\phi(N_i) \cap (\{m\} \times \mathbb{N})$ is finite for all of those $i \in F$, thus any $j \in S'_{Y,r+1} \setminus \cup_{i \in F} N_i$ would violate the conclusion of Lemma 6.1. In particular for those i such that $\phi(i) \in (\{\phi(i')^1\} \times \mathbb{N})$, $i' \in I$, $w_{\phi(i)} = w_{\phi(i')}$ and thus only finitely many of these N_i will fail to contain a subset N'_i as above. Similarly, if $F \subset \{i : N_i \not\supset N'_i \ni J \cup \{j_{r+1}\} \subset N'_i, N'_i \in \mathcal{S}(Y) \text{ but } \exists N \subset N_i, N \in S(Y)\}$ and $\sum_{i \in F} w_{\phi(i)}^{2p/(p-2)} > rH(\epsilon_r, \min\{w_{\phi(i')} : i' \in I\}, r)$. then $\phi(N_i) \cap (\{\phi(j)^1\} \times \mathbb{N})$ is finite for some $j \in J \cup \{j_{r+1}\}$ for all of those $i \in F$, thus for at least one of $j' = j_1, j_2, \ldots, j_{r+1}$ there exists $F' \subset F$ such that $\sum_{i \in F'} w_{\phi(i)}^{2p/(p-2)} > H(\epsilon_r, \min\{w_{\phi(i')} : i' \in I\}, r)$ and $j \in S'_{Y,r+1} \setminus \cup_{i \in F} N_i$ with $\phi(j)^1 = \phi(j')$. This would violate the conclusion of Lemma 6.1. It follows that the required set $S_{X,r+1}$ exists.

Player r+1 selects i_{r+1} from $S_{X,r+1}$ and selects $S'_{X,r+1} \subset S_{X,r+1}$ which is in $\mathcal{S}(X)$ and contains $\{i_1, i_2, \ldots, i_{r+1}\}$. By our choice of $S_{X,r+1}$, $N_{i_{r+1}}$ contains $N'_{i_{r+1}}$ such that $\{j_1, \ldots, j_{r+1}\} \subset N'_{i_{r+1}} \in \mathcal{S}(Y)$. The set $S = N'_{i_{r+1}}$ is our candidate for $S_{Y,r+2}$.

This completes the induction step.

Therefore there are subsets $I = \{i_1, i_2, \ldots\}$, $J = \{j_1, j_2, \ldots\}$ of $\mathbb{N}$ with $\phi(I)$ and $\phi(J)$ rich such that

(1) $|(x^*_{\phi(i_s)} \otimes y^*_{\phi(j)})(Tx_{\phi(i_t)} \otimes y_{\phi(j)})| < \epsilon_{(s \vee t)-1} w_{\phi(i_t)}/((s \vee t)-1)$ for all $s \neq t \in \mathbb{N}$ and $j \in J$.

(2) $|(x^*_{\phi(i)} \otimes y^*_{\phi(j_s)})(Tx_{\phi(i)} \otimes y_{\phi(j_t)})| < \epsilon_{(s \vee t)-1} w_{\phi(j_t)}/((s \vee t)-1)$ for all $i \in I$ and $s \neq t \in \mathbb{N}$

(3) $(x^*_{\phi(i)} \otimes y^*_{\phi(j)})(Tx_{\phi(i')} \otimes y_{\phi(j')}) = 0$ for all $i' \in I$ and $j' \in J$ such that $i \neq i'$ and $j \neq j'$; $i = i'$ and $o(j \vee j') > o(i), j \neq j'$; or $i \neq i', o(i \vee i') \geq o(j)$ and $j = j'$.

It remains to show that the restriction of T to $Z = [x_{\phi(i)} \otimes y_{\phi(j)} : i \in I, j \in J]$ is an isomorphism. As noted above we may assume by composing with the projection onto Z that T is actually a map from Z to itself and that $\epsilon = 1$.

Moreover, we can assume that $I = J = \mathbb{N}$ by replacing i by $o(i, I)$ and j by $o(j, J)$. Thus with the notation appropriately revised we have

(0) $(x^*_{\phi(i)} \otimes y^*_{\phi(j)})(Tx_{\phi(i)} \otimes y_{\phi(j)}) = 1$ for all i, j,

(1) $|(x^*_{\phi(i)} \otimes y^*_{\phi(j)})(Tx_{\phi(i')} \otimes y_{\phi(j)})| < \epsilon_{i \vee i' - 1} w_{\phi(i')}/(i \vee i' - 1)$ for all $i \neq i' \in \mathbb{N}$ and $j \in \mathbb{N}$.

(2) $|(x^*_{\phi(i)} \otimes y^*_{\phi(j)})(Tx_{\phi(i)} \otimes y_{\phi(j')})| < \epsilon_{j \vee j' - 1} w_{\phi(j')}/(j \vee j' - 1)$ for all $i \in \mathbb{N}$ and $j \neq j' \in \mathbb{N}$

(3) $(x^*_{\phi(i)} \otimes y^*_{\phi(j)})(Tx_{\phi(i')} \otimes y_{\phi(j')}) = 0$ for all $i' \in I$ and $j' \in \mathbb{N}$ such that $i \neq i'$ and $j \neq j'$; $i = i'$ and $j \vee j' > i, j \neq j'$; or $i \neq i', i \vee i' \geq j$ and $j = j'$.

We will now estimate $\|Tz - z\|$ for $z \in Z$ with finite support relative to the basis.

Let $z = \sum_{i \in \mathbb{N}} \sum_{j \in \mathbb{N}} a_{i,j} x_{\phi(i)} \otimes y_{\phi(j)})$.

First we estimate the ℓ_2-norm.

$\|Tz - z\|_2^2$

$$= \sum_{i \in \mathbb{N}} \sum_{j \in \mathbb{N}} \left| \sum_{\substack{i',j' \in \mathbb{N} \\ (i',j') \neq (i,j)}} a_{i',j'} (x^*_{\phi(i)} \otimes y^*_{\phi(j)})(Tx_{\phi(i')} \otimes y_{\phi(j')}) \right|^2 w^2_{\phi(i)} w^2_{\phi(j)}$$

$$= \sum_{j=1}^{\infty} \sum_{i<j} \left| \sum_{\substack{i',j' \in \mathbb{N} \\ (i',j') \neq (i,j)}} a_{i',j'} (x^*_{\phi(i)} \otimes y^*_{\phi(j)})(Tx_{\phi(i')} \otimes y_{\phi(j')}) \right|^2 w^2_{\phi(i)} w^2_{\phi(j)}$$

$$+ \sum_{i=1}^{\infty} \sum_{j \leq i} \left| \sum_{\substack{i',j' \in \mathbb{N} \\ (i',j') \neq (i,j)}} a_{i',j'} (x^*_{\phi(i)} \otimes y^*_{\phi(j)})(Tx_{\phi(i')} \otimes y_{\phi(j')}) \right|^2 w^2_{\phi(i)} w^2_{\phi(j)}$$

$$= \sum_{j=1}^{\infty} \sum_{i<j} \left| \sum_{1 \leq i' \neq i < j} a_{i',j} (x^*_{\phi(i)} \otimes y^*_{\phi(j)})(Tx_{\phi(i')} \otimes y_{\phi(j)}) \right|^2 w^2_{\phi(i)} w^2_{\phi(j)}$$

$$+ \sum_{i=1}^{\infty} \sum_{j \leq i} \left| \sum_{1 \leq j' \neq j \leq i} a_{i',j'} (x^*_{\phi(i)} \otimes y^*_{\phi(j)})(Tx_{\phi(i)} \otimes y_{\phi(j')}) \right|^2 w^2_{\phi(i)} w^2_{\phi(j)}$$

by (3)

$$\leq \sum_{j=1}^{\infty} \sum_{i<j} \left(\sum_{1 \leq i' < i} |a_{i',j}| \frac{\epsilon_{i-1} w_{\phi(i')}}{i-1} + \sum_{i' : i < i' < j} |a_{i',j}| \frac{\epsilon_{i'-1} w_{\phi(i')}}{i'-1} \right)^2 w^2_{\phi(i)} w^2_{\phi(j)}$$

$$+ \sum_{i=1}^{\infty} \sum_{j \leq i} \left(\sum_{1 \leq j' < j} |a_{i,j'}| \frac{\epsilon_{j-1} w_{\phi(j')}}{j-1} j + \sum_{j < j' \leq i} |a_{i,j'}| \frac{\epsilon_{j'-1} w_{\phi(j')}}{j'-1} \right)^2 w^2_{\phi(i)} w^2_{\phi(j)}$$

by (1) and (2)

$$\leq \sum_{j=1}^{\infty} \sum_{i<j} \left(\sum_{1 \leq i' < i} |a_{i',j}|^2 \frac{\epsilon_{i-1} w_{\phi(i')}^2}{i-1} + \sum_{i':i<i'<j} |a_{i',j}|^2 \frac{\epsilon_{i'-1} w_{\phi(i')}^2}{i'-1} \right) w_{\phi(i)}^2 w_{\phi(j)}^2$$

$$+ \sum_{i=1}^{\infty} \sum_{j \leq i} \left(\sum_{1 \leq j' < j} |a_{i,j'}|^2 \frac{\epsilon_{j-1} w_{\phi(j')}^2}{j-1} + \sum_{j<j' \leq i} |a_{i,j'}|^2 \frac{\epsilon_{j'-1} w_{\phi(j')}^2}{j'-1} \right) w_{\phi(i)}^2 w_{\phi(j)}^2$$

by the convexity of t^2

$$\leq \sum_{j=1}^{\infty} \sum_{i'=1}^{j-1} |a_{i',j}|^2 w_{\phi(i')}^2 w_{\phi(j)}^2 \Big(\sum_{i:j>i>i'} \frac{w_{\phi(i)}^2 \epsilon_{i-1}}{i-1} + \sum_{i:i<i'} \frac{\epsilon_{i'-1} w_{\phi(i)}^2}{i'-1} \Big)$$

$$+ \sum_{i=1}^{\infty} \sum_{j'=1}^{i} |a_{i,j'}|^2 w_{\phi(i)}^2 w_{\phi(j')}^2 \Big(\sum_{j:i \geq j>j'} \frac{w_{\phi(j)}^2 \epsilon_{j-1}}{j-1} + \sum_{j:j<j'} \frac{\epsilon_{j'-1} w_{\phi(j)}^2}{j'-1} \Big)$$

$$\leq \|z\|_2^2 \epsilon' / 4^p$$

Next we estimate the ℓ_p-norm.

$$|Tz - z|_p^p = \sum_{i \in \mathbb{N}} \sum_{j \in \mathbb{N}} \left| \sum_{\substack{i',j' \in \mathbb{N} \\ (i',j') \neq (i,j)}} a_{i',j'} (x_{\phi(i)}^* \otimes y_{\phi(j)}^*)(T x_{\phi(i')} \otimes y_{\phi(j')}) \right|^p$$

$$= \sum_{j=1}^{\infty} \sum_{i<j} \left| \sum_{1 \leq i' \neq i < j} a_{i',j} (x_{\phi(i)}^* \otimes y_{\phi(j)}^*)(T x_{\phi(i')} \otimes y_{\phi(j)}) \right|^p$$

$$+ \sum_{i=1}^{\infty} \sum_{j \leq i} \left| \sum_{1 \leq j' \neq j \leq i} a_{i',j'} (x_{\phi(i)}^* \otimes y_{\phi(j)}^*)(T x_{\phi(i)} \otimes y_{\phi(j')}) \right|^p$$

(by (3))

$$\leq \sum_{j=1}^{\infty} \sum_{i<j} \left(\sum_{1 \leq i' < i} |a_{i',j}| \frac{\epsilon_{i-1} w_{\phi(i')}}{i-1} + \sum_{i':i<i'<j} |a_{i',j}| \frac{\epsilon_{i'-1} w_{\phi(i')}}{i'-1} \right)^p$$

$$+ \sum_{i=1}^{\infty} \sum_{j \leq i} \left(\sum_{1 \leq j' < j} |a_{i,j'}| \frac{\epsilon_{j-1} w_{\phi(j')}}{j-1} + \sum_{j<j' \leq i} |a_{i,j'}| \frac{\epsilon_{j'-1} w_{\phi(j')}}{j'-1} \right)^p$$

(by (1) and (2))

$$\leq \sum_{j=1}^{\infty} \sum_{i<j} \left(\sum_{1 \leq i' < i} |a_{i',j}|^p \frac{\epsilon_i w_{\phi(i')}}{i-1} + \sum_{i':i<i'<j} |a_{i',j}|^p \frac{\epsilon_{i'-1} w_{\phi(i')}}{i'-1} \right)$$

$$+ \sum_{i=1}^{\infty} \sum_{j \leq i} \left(\sum_{1 \leq j' < j} |a_{i,j'}|^p \frac{\epsilon_{j-1} w_{\phi(j')}}{j-1} + \sum_{j<j' \leq i} |a_{i,j'}|^p \frac{\epsilon_{j'-1} w_{\phi(j')}}{j'-1} \right)$$

(by the convexity of t^p)

$$\leq \sum_{j=1}^{\infty} \sum_{i':1\leq i'<j} |a_{i',j}|^p \Big(\sum_{i:i>i'} \frac{w_{\phi(i)}\epsilon_{i-1}}{i-1} + \sum_{i:i<i'} \frac{\epsilon_{i'-1}w_{\phi(i)}}{i'-1} \Big)$$

$$+ \sum_{i=1}^{\infty} \sum_{j':1\leq j'<i} |a_{i,j'}|^p \Big(\sum_{j:j>j'} \frac{w_{\phi(j)}\epsilon_{j-1}}{j-1} + \sum_{j:j<j'} \frac{\epsilon_{j'-1}w_{\phi(j)}}{j'-1} \Big)$$

$$\leq |z|_p^p \epsilon'/4^p$$

The estimates for the two row and column norms are similar so we will only do one.

$$\|Tz-z\|_R^p$$

$$= \sum_{i\in\mathbb{N}} \left(\sum_{j\in\mathbb{N}} \left| \sum_{\substack{i',j'\in\mathbb{N}\\(i',j')\neq(i,j)}} a_{i',j'}(x_{\phi(i)}^*\otimes y_{\phi(j)}^*)(Tx_{\phi(i')}\otimes y_{\phi(j')}) \right|^2 w_{\phi(j)}^2 \right)^{\frac{p}{2}}$$

$$\leq 2^{p-1}\left(\sum_{i=1}^{\infty}\left(\sum_{j>i}\left|\sum_{\substack{i',j'\in\mathbb{N}\\(i',j')\neq(i,j)}} a_{i',j'}(x_{\phi(i)}^*\otimes y_{\phi(j)}^*)(Tx_{\phi(i')}\otimes y_{\phi(j')})\right|^2 w_{\phi(j)}^2\right)^{\frac{p}{2}} \right.$$

$$\left. + \sum_{i=1}^{\infty}\left(\sum_{j\leq i}\left|\sum_{\substack{i',j'\in\mathbb{N}\\(i',j')\neq(i,j)}} a_{i',j'}(x_{\phi(i)}^*\otimes y_{\phi(j)}^*)(Tx_{\phi(i')}\otimes y_{\phi(j')})\right|^2 w_{\phi(j)}^2\right)^{\frac{p}{2}} \right)$$

$$= 2^{p-1}\left(\sum_{i=1}^{\infty}\left(\sum_{j>i}\left|\sum_{1\leq i'\neq i<j} a_{i',j}(x_{\phi(i)}^*\otimes y_{\phi(j)}^*)(Tx_{\phi(i')}\otimes y_{\phi(j)})\right|^2 w_{\phi(j)}^2\right)^{\frac{p}{2}} \right.$$

$$\left. + \sum_{i=1}^{\infty}\left(\sum_{j\leq i}\left|\sum_{1\leq j'\neq j\leq i} a_{i',j'}(x_{\phi(i)}^*\otimes y_{\phi(j)}^*)(Tx_{\phi(i)}\otimes y_{\phi(j')})\right|^2 w_{\phi(j)}^2\right)^{\frac{p}{2}} \right)$$

(by (3))

$$\leq 2^{p-1}\left(\sum_{i=1}^{\infty}\left(\sum_{j>i}\left(\sum_{i'=1}^{i-1}|a_{i',j}|\frac{\epsilon_{i-1}w_{\phi(i')}}{(i-1)} + \sum_{i'=i+1}^{j-1}|a_{i',j}|\frac{\epsilon_{i'-1}w_{\phi(i')}}{(i'-1)}\right)^2 w_{\phi(j)}^2\right)^{\frac{p}{2}} \right.$$

$$\left. + \sum_{i=1}^{\infty}\left(\sum_{j\leq i}\left(\sum_{j'=1}^{j-1}|a_{i,j'}|\frac{\epsilon_{j-1}w_{\phi(j')}}{j-1} + \sum_{j'=j+1}^{i}|a_{i,j'}|\frac{\epsilon_{j'-1}w_{\phi(j')}}{(j'-1)}\right)^2 w_{\phi(j)}^2\right)^{\frac{p}{2}} \right)$$

(by (1) and (2))

$$\leq 2^{p-1} \left(\sum_{i=1}^{\infty} \left(\sum_{j>i} \left(\sum_{i'=1}^{i-1} |a_{i',j}|^2 \frac{\epsilon_{i-1} w_{\phi(i')}^2}{(i-1)} + \sum_{i'=i+1}^{j-1} |a_{i',j}|^2 \frac{\epsilon_{i'-1} w_{\phi(i')}^2}{i'-1} \right) w_{\phi(j)}^2 \right)^{\frac{p}{2}} \right.$$

$$\left. + \sum_{i=1}^{\infty} \left(\sum_{j\leq i} \left(\sum_{j'=1}^{j-1} |a_{i,j'}|^2 \frac{\epsilon_{j-1} w_{\phi(j')}^2}{j-1} + \sum_{j<j'\leq i} |a_{i,j'}|^2 \frac{\epsilon_{j'-1} w_{\phi(j')}^2}{j'-1} \right) w_{\phi(j)}^2 \right)^{\frac{p}{2}} \right)$$

(by the convexity of t^2)

$$\leq 2^{p-1} \left(\sum_{i=1}^{\infty} \left(\sum_{i':1\leq i'<i} \left(\sum_{j>i} |a_{i',j}|^2 w_{\phi(j)}^2 \right)^{\frac{p}{2}} \frac{w_{\phi(i')}^2 \epsilon_{i-1}}{i-1} \right. \right.$$

$$\left. + \sum_{i'>i} \left(\sum_{j>i'} |a_{i',j}|^2 w_{\phi(j)}^2 \right)^{\frac{p}{2}} \frac{\epsilon_{i'-1} w_{\phi(i)}^2}{i'-1} \right)$$

$$\left. + \sum_{i=1}^{\infty} \left(\sum_{j':1\leq j'\leq i} |a_{i,j'}|^2 w_{\phi(j')}^2 \left(\sum_{j:j>j'} \frac{w_{\phi(j)}^2 \epsilon_{j-1}}{j-1} + \sum_{j:j<j'} \frac{\epsilon_{j'-1} w_{\phi(j)}^2}{j'-1} \right) \right)^{\frac{p}{2}} \right)$$

(by the convexity of $t^{\frac{p}{2}}$)

$$\leq 2^{p-1} \left(\sum_{i=1}^{\infty} \left(\sum_{j>i} |a_{i',j}|^2 w_{\phi(j)}^2 \right)^{\frac{p}{2}} \frac{\epsilon'}{4^p} \right.$$

$$\left. + \sum_{i=1}^{\infty} \left(\sum_{j':1\leq j'\leq i} |a_{i,j'}|^2 w_{\phi(j')}^2 \right)^{\frac{p}{2}} \left(\frac{\epsilon'}{4^p} \right)^{\frac{p}{2}} \right)$$

$$\leq \|z\|_R^p \frac{2^p \epsilon'}{4^p}$$

Because $\epsilon' < 1$, we have that there is an $c < 1$ such that $\|Tz - z\| < c\|z\|$ for all $z \in Z$. Thus $T|_Z$ is an isomorphism. $\square$

Notice that in fact we can make the constant c in the proof above as close to 0 as we wish.

Proposition 6.3 shows that we can focus on the diagonal coordinates of an operator on $X_p \otimes X_p$ in attempting to show that an operator is large. We will use this in Section 8 to show that certain operators do not exist. For further development of the isomorphic theory it is also necessary to know that there are many natural operators on a space. We will finish this section by looking at some of the natural classes of operators and some interesting subsets of the natural bases of $X_p \otimes X_p$.

DEFINITION 6.4. If T is an operator from $X_{p,(w_n)} \otimes X_{p,(w'_n)}$ into $X_{p,(u_n)} \otimes X_{p,(u'_n)}$, say that T is $(p, R, C, 2)-$bounded if it is bounded with respect to each of the four norms, i.e., there is a constant K such that for all $z \in X_{p,(w_n)} \otimes X_{p,(w'_n)}$, $\|Tz\|_p \le K\|z\|_p, \|Tz\|_R \le K\|z\|_R, \|Tz\|_C \le K\|z\|_C$ and $\|Tz\|_2 \le K\|z\|_2$.

Observe that if T_1 is a $(p, 2)-$bounded operator from $X_{p,(w_n)}$ into $X_{p,(u_n)}$ and T_2 is a $(p, 2)-$bounded operator from $X_{p,(w'_n)}$ into $X_{p,(u'_n)}$, then $T_1 \otimes T_2$ is $(p, R, C, 2)-$bounded. Also any basis projection relative to the natural basis of $X_{p,(w_n)} \otimes X_{p,(w'_n)}$ is $(p, R, C, 2)-$bounded. Thus there are many such operators.

LEMMA 6.5. *If (F_i) and (G_i) are sequences of finite subsets of $\mathbb{N}$ such that* $\max F_i < \min F_{i+1}$ *and* $\max G_i < \min G_{i+1}$ *then*

$$[e_n \otimes e'_m : (n, m) \in F_i \times G_i \text{ for some } i],$$

where (e_n) and (e'_m) are natural bases for $X_{p,(w_n)}$ and $X_{p,(w'_n)}$, respectively, is isomorphic to a complemented subspace of X_p.

Proof. Observe that the block subspaces $Z_i = [e_n \otimes e'_m : (n, m) \in F_i \times G_i]$ are $(p, 2)-$complemented in $X_{p,(w_n)} \otimes X_{p,(w'_n)}$ and that $Z = \sum_{i \in \mathbb{N}} Z_i$ is a natural $(p, 2)-$sum decomposition. Therefore, Z is $(p, 2)-$isomorphic to a $(p, 2)-$complemented subspace of R_p^ω and thus of X_p.

It follows from Lemma 6.5 and the fact that if Y is a complemented subspace of X_p then $Y \oplus X_p$ is isomorphic to X_p, [JO], that $X_{p,(w_n)} \otimes X_{p,(w'_n)}$ is isomorphic to $[e_n \otimes e'_m : n \ne m]$. Thus the diagonal is a negligible subspace of $X_p \otimes X_p$. We will next show that the upper (lower) triangle of $X_p \otimes X_p$ is isomorphic to the whole space. Notice that this question depends on the representation of X_p. There are some representations where the argument follows from a Cantor-Bernstein result [W2], [Woj, Corollary 4.6].

LEMMA 6.6. *Suppose that (w_n) and (w'_n) are sequences in $(0, 1]$ which satisfy (*) and that $w_{n,k} = w_n$ and $w'_{n,k} = w'_n$ for all $n \in \mathbb{N}$. Then $X_{p,(w_{n,k})} \otimes X_{p,(w'_{n,k})}$ is isomorphic to $[e_{\phi(i)} \otimes e_{\phi(j)} : i < j]$, where $\phi : \mathbb{N} \to \mathbb{N} \times \mathbb{N}$ is a bijection as in Section 5.*

Proof. It is sufficient to show that $(e_{\phi(i)} \otimes e_{\phi(j)})_{i<j}$ contains a subsequence which is equivalent to the natural basis of $X_{p,(w_{n,k})} \otimes X_{p,(w'_{n,k})}$. Observe that the reflection along the diagonal mapping, R, defined by $R(e_{\phi(i)} \otimes e_{\phi(j)}) = e_{\phi(j)} \otimes e_{\phi(i)}$ extends linearly to an isomorphism. Because every weight is repeated infinitely often, we can split the columns with weight w_n into two infinite sets with indices K_n and L_n, i.e., for $k \in K_n \cup L_n$, $\|e_{\phi(k)}\|_2 = w_n$, $K_n \cap L_n = \emptyset$ and $|K_n| = |L_n| = \infty$. Let $\psi_n : K_n \cup L_n \to K_n$ be injective and satisfy $\psi_n(k) > k$ for all k and let $\zeta_n : K_n \cup L_n \to L_n$ do likewise. Define $\psi : \mathbb{N} \to \cup_n K_n$ by $\psi(i) = \psi_n(i)$ if $i \in K_n \cup L_n$ and $\zeta : \mathbb{N} \to \cup_n L_n$ by $\zeta(i) = \zeta_n(i)$ if $i \in K_n \cup L_n$. Define

$$S(e_{\phi(i)} \otimes e_{\phi(j)}) = \begin{cases} e_{\phi(i)} \otimes e_{\phi(\psi(j))}, & \text{if } i \le j \\ e_{\phi(j)} \otimes e_{\phi(\zeta(i))}, & \text{if } j < i, \end{cases}$$

and extend linearly. It is easy to check that S is an isomorphism. $\square$

LEMMA 6.7. *Suppose that (w_n) and (w'_n) are sequences in $(0,1]$ which satisfy (*) and that $w_{n,k} = w_n$ and $w'_{n,k} = w'_n$ for all $n, k \in \mathbb{N}$. Let (u_n) and (u'_n) be two sequences in $(0,1]$ which satisfy (*). Let $(e_{n,k})$, $(e'_{n,k})$, (d_n) and (d'_n) be the natural bases for $X_{p,(w_{n,k})}$, $X_{p,(w'_{n,k})}$, $X_{p,(u_n)}$ and $X_{p,(u'_n)}$, respectively. Then there is a $(p, R, C, 2)-$ complemented subspace of $[d_n \otimes d'_m : n < m]$ which is $(p, R, C, 2)-$ isomorphic to $[e_{\phi(i)} \otimes e_{\phi(j)} : i < j]$.*

Proof. By [R] there is a block basic sequence (D_n) of the basis, (d_n), of $X_{p,(u_n)}$ which is $(p, 2)-$ equivalent to the basis $(e_{n,k})$ with $(p, 2)-$ complemented closed span. Similarly, there is a block basic sequence (D'_n) of the basis, (d'_n), of $X_{p,(u'_n)}$ which is $(p, 2)-$ equivalent to the basis $(e'_{n,k})$ with $(p, 2)-$ complemented closed span. Let P and P' be the corresponding projections. Then $P \otimes P'$ is a $(p, R, C, 2)-$ bounded projection from $X_{p,(u_n)} \otimes X_{p,(u'_n)}$ onto the subspace $[D_n \otimes D'_m]$. If we restrict this map to $[d_n \otimes d'_m : n < m]$ and compose with a basis projection to eliminate any partial support of blocks $D_n \otimes D'_n$, we get the required projection and subspace. $\square$

PROPOSITION 6.8. *Let (u_n) and (u'_n) be two sequences in $(0,1]$ which satisfy (*) and let (d_n) and (d'_n) be the natural bases for $X_{p,(u_n)}$ and $X_{p,(u'_n)}$, respectively. Then $[d_n \otimes d'_m : n < m]$ is isomorphic to $X_p \otimes X_p$.*

Proof. It follows from Lemma 6.7 with $w_n = u_n$ and $w'_n = u'_n$ and an argument analogous to that of Rosenthal [R, Theorem 13] that $[d_n \otimes d'_m : n < m]$ is isomorphic to its square and to $[e_{\phi(i)} \otimes e'_{\phi(j)} : i < j]$, with the same notation as in the previous lemma. By Lemma 6.6 $[e_{\phi(i)} \otimes e'_{\phi(j)} : i < j]$ is isomorphic to $X_p \otimes X_p$. $\square$

CHAPTER 7

ISOMORPHISMS OF $X_p \otimes X_p$ ONTO
COMPLEMENTED SUBSPACES OF $(p,2)-$SUMS

Our goal is to show that $X_p \otimes X_p$ is not isomorphic to a complemented sub-space of any $R_p^\alpha, \alpha < \omega_1$. Because $X_p \otimes X_p$ is isomorphic to a subspace of $R_p^{\omega 2}$ the projection must play a fundamental role. We have already seen that R_p^α is isomorphic to a $(p,2)-$sum of spaces $R_p^{\alpha_n}$. In this section we will develop a number of technical results which describe the restrictions on a complemented isomorphic embedding of $X_p \otimes X_p$ into a $(p,2)-$sum of subspaces of L_p.

Below we will be working with a $(p,2)-$sum of spaces Y_j and we will denote the natural projection onto $[Y_j : j \leq k]$ by P_k. We will use the sequence space norm rather than the norm as an independent sum. In applications to R_p^α we will need to change to the independent sum and this will introduce a constant C, the constant in Rosenthal's inequality, from the equivalence between the norm on $(\sum Y_j)_{p,2,(w_n)}$ and and embedding Y_n into L_p on independent coordinates and using the L_p-norm. This change is only an annoying technicality.

The proof of the next lemma and the one following are somewhat easier if we assume that each space Y_n has an unconditional basis and that we have done a quasi-blocking of the image of the basis of $X_p \otimes X_p$ relative to the unconditional basis of $(\sum Y_n)_{p,2}$ as in the conclusion of Proposition 5.4. In that case we can arrange things so that some of the error estimates can be replaced by zero and get a stronger conclusion. We summarize this as Lemma 7.4 below, but the reader may find the proofs of the first two lemmas are easier to understand on the first reading if he considers this easier case.

The first lemma is closely related to the results of Section 4 except that we assume that we have a projection onto the range of the operator. Throughout this section and the next we will assume that the operator from $X_p \otimes X_p$ satisfies (T2). There is no loss of generality since by Proposition 4.2 we can always restrict the operator to a natural complemented subspace which is isomorphic to the whole space and so that the restricted operator satisfies (T2).

LEMMA 7.1. *Let T be an isomorphism from $X_{p,(w_n)} \otimes X_{p,(w_n)}$ into $(\sum Y_j)_{p,2,(w_n')}$ satisfying (T2) and let P be a projection onto the range of T. Let ρ, ρ' and δ be positive constants with $\rho > \rho'$ and let (x_i) and (y_j) be two copies of the usual basis for $X_{p,(w_n)}$. Suppose that j is such that $w_j < \frac{\delta^{1/2}(3\rho+\rho')}{6\sqrt{2}\|T\|\|T^{-1}\|\|P\|}$, $(N_k)_{k=1}^K$, and $(N_k')_{k=1}^K$ are strictly increasing sequences of integers with $N_k \leq N_k' < N_{k+1}$ for all k, $(F_k)_{k=1}^K$ is a disjoint sequence of finite subsets of $\mathbb{N}$ such* *that for all $i \in F_k$*

(7.1.1)
$$|x_i^* \otimes y_j^*(T^{-1}P(P_{N_k} - P_{N_{k-1}'})T(x_i \otimes y_j))| \geq \rho,$$

(7.1.2)
$$|x_i^* \otimes y_j^*(T^{-1}P(P_{N_k} - P_{N_{k-1}'}) \sum_{\substack{s \neq i \\ s \in F_k}} w_s^{2/(p-2)} T(x_s \otimes y_j))| \leq (\rho - \rho') w_i^{2/(p-2)}/4$$

and for all k

(7.1.3)
$$1 \geq \sum_{i \in F_k} w_i^{2p/(p-2)} \geq \delta$$

Then $K \leq \max\{2\delta^{-1}, (\|T\|^2\|T^{-1}\|\|P\|8/(\delta^{1/2}(3\rho + \rho')))^{2p/(p-2)}\}$.

Proof. First by replacing ρ by $\rho/2$ we may assume that for all $i \in F_k$,

$$x_i^* \otimes y_j^*(T^{-1}P(P_{N_k} - P_{N_{k-1}'})T(x_i \otimes y_j)) \geq \rho,$$

that is, they all have the same sign which we assume to be positive. Because $(r_i T x_i \otimes y_j)_{i \in F_k}$, where (r_i) is the sequence of Rademacher, is orthogonal in $L_2(\Omega \times [0,1])$, there is a choice of signs (ϵ_i) such that

(7.1.4)
$$\|\sum_{i \in F_k} \epsilon_i w_i^{2/(p-2)} T x_i \otimes y_j\|_2 \leq \frac{3}{2}(\sum_{i \in F_k} w_i^{4/(p-2)} \|T x_i \otimes y_j\|_2^2)^{1/2}.$$

Let $z_k = \sum_{i \in F_k} \epsilon_i w_i^{2/(p-2)} x_i \otimes y_j$ for each k. Since $(x_i \otimes y_j)_i$ is equivalent to the usual basis of $X_{p,(w_i)}$ it follows from Proposition 0.2 that $[z_k]$ is complemented in $[x_i \otimes y_j : i \in \mathbb{N}]$ with projection Q_1. This in turn induces an operator Q from $(\sum Y_j)_{p,2,(w_n)}$ onto $[z_k]$, namely, $Q = Q_1 T^{-1} P$. Since we will need to do some computations with Q, let us be more explicit about its evaluation. Let

$$z_k^* = (\sum_{i \in F_k} w_i^{2p/(p-2)})^{-1} \sum_{i \in F_k} \epsilon_i w_i^{2(p-1)/(p-2)} x_i^* \otimes y_j^*.$$

Then $Q(z) = \sum_{k=1}^K z_k^*(T^{-1}Pz)z_k$.

Let $R_k = (P_{N_k} - P_{N_{k-1}'})$ for all k. Now we want to restrict Q to the subspace $Z = [R_k T(z_k)]$ and pass to its "diagonal". Observe that

$$(\sum_{i \in F_k} w_i^{2p/(p-2)})|z_k^*(T^{-1}PR_kT(z_k))|$$

$$= |\sum_{i \in F_k}[w_i^{2p/(p-2)}(x_i^* \otimes y_j^*)(T^{-1}PR_kT(x_i \otimes y_j))$$

$$+ w_i^{2(p-1)/(p-2)}(x_i^* \otimes y_j^*)(T^{-1}PR_k \sum_{\substack{s \neq i \\ s \in F_k}} w_s^{2/(p-2)} T(x_s \otimes y_j))]|$$

$$\geq \sum_{i \in F_k} w_i^{2p/(p-2)}\rho - w_i^{2(p-1)/(p-2)}w_i^{2/(p-2)}(\rho - \rho')/4$$

(by 7.1.2)

$$(7.1.5) \qquad = \sum_{i \in F_k} w_i^{2p/(p-2)} (3\rho + \rho')/4.$$

Therefore $\|R_k T z_k\| \geq (3\rho + \rho')/(4\|T\|\|P\|)$. Thus we have that $(R_k T(z_k))$ is a sequence of non-zero blocks in an unconditional sum with constant 1 and thus is an unconditional basic sequence. Applying Tong's Theorem [L-T, Proposition 1.c.8] to Q restricted to the subspace Z, we get that the operator Q_D, defined by

$$Q_D(\sum a_k R_k T z_k) = \sum a_k z_k^*(T^{-1} P R_k T z_k) z_k,$$

is bounded with the norm at most $\|T\|\|T^{-1}\|\|P\|$.

Now let us estimate the norm more directly. First because R_k is contractive in both norms.

$$\|\sum a_k R_k T z_k\| \leq \max\{(\sum |a_k|^p \|R_k T z_k\|_p^p)^{1/p}, (\sum |a_k|^2 \|R_k T z_k\|_2^2)^{1/2}\}$$

$$\leq \max\{(\sum |a_k|^p \|T z_k\|_p^p)^{1/p},$$

$$(\sum |a_k|^2 \frac{9}{4} \sum_{i \in F_k} w_i^{4/(p-2)} \|T x_i \otimes y_j\|_2^2)^{1/2}\}$$

(by 7.1.4)

$$\leq \max\{(\sum |a_k|^p \sum_{i \in F_k} w_i^{2p/(p-2)})^{1/p} \|T\|,$$

$$(\sum |a_k|^2 \frac{9}{4} \sum_{i \in F_k} w_i^{2p/(p-2)} w_j^2 \|T\|^2)^{1/2}\}$$

(by (T2))

$$(7.1.6) \qquad \leq \|T\| \max\{(\sum |a_k|^p)^{1/p}, (\frac{3}{2}) w_j (\sum |a_k|^2)^{1/2}\}$$

since $\sum_{i \in F_k} w_i^{2p/(p-2)} \leq 1$.

Next we estimate the norm of the image under Q_D. We consider only the case $a_k = 1$ for all k. Let $z_k' = T^{-1} P R_k T z_k$ for all k.

$$\|\sum_k z_k^*(z_k') z_k\| \geq \max\{(\sum_k |z_k^*(z_k')|^p \sum_{i \in F_k} w_i^{2p/(p-2)})^{1/p},$$

$$(\sum_k |z_k^*(z_k')|^2 \sum_{i \in F_k} w_i^{2p/(p-2)})^{1/2}\}$$

$$\geq ((3\rho + \rho')/4) \max\{(\sum_k \sum_{i \in F_k} w_i^{2p/(p-2)})^{1/p},$$

$$(\sum_k \sum_{i \in F_k} w_i^{2p/(p-2)})^{1/2}\}$$

(by 7.1.5)

$$(7.1.7) \qquad \geq ((\rho + 3\rho')/4) \max\{(\delta/2)^{1/p} K^{1/p}, (\delta/2)^{1/2} K^{1/2}\}$$

Combining the estimates 7.1.6 and 7.1.7 yields,

$$\|Q_D\| \|T\| \max\{K^{1/p}, (\tfrac{3}{2}) w_j K^{1/2}\} \geq ((\rho + 3\rho')/4) \max\{(\delta K/2)^{1/p}, (\delta K/2)^{1/2}\}.$$

If K is such that $K\delta \geq 2$, then the inequality becomes

$$\|T^{-1}\| \|P\| \|T\|^2 \max\{K^{1/p}, (\tfrac{3}{2}) w_j K^{1/2}\} \geq ((3\rho + \rho')/4)(K\delta/2)^{1/2},$$

and thus either

$$K^{1/p} < (\tfrac{3}{2}) w_j K^{1/2}$$

from which it would follow that

$$\|T^{-1}\| \|P\| \|T\|^2 (6\sqrt{2}) w_j > (3\rho + \rho') \delta^{1/2},$$

or

$$\sqrt{2} \|T^{-1}\| \|P\| \|T\|^2 \geq ((3\rho + \rho')/4)(\delta)^{1/2} K^{1/2 - 1/p}.$$

Because we have assumed that w_j is small, the former is impossible and the conclusion follows. $\square$

The next lemma is essentially a reformulation of Lemma 7.1 for the special case of the weights $(w_{n,k})$.

LEMMA 7.2. *Suppose that (x_i) and (y_j) are bases of $X_{p,(w_{n,k})}$. Let T be an isomorphism from $X_{p,(w_{n,k})} \otimes X_{p,(w_{n,k})}$ into $(\sum Y_j)_{p,2,(w_n)}$ satisfying (T2) and let P be a projection onto the range of T. Let ρ, ρ_1 and δ be positive constants, $\rho > \rho_1$. Then there exist integers M_m, $m \in \mathbb{N}$ such that if $w_j < \frac{\delta^{1/2}(4\rho - \rho_1)}{6\sqrt{2}\|T\| \|T^{-1}\| \|P\|}$, $(N_k)_{k=1}^K$ and $(N_k')_{k=1}^K$ are strictly increasing sequences of integers with $N_k \leq N_k' \leq N_{k+1}$ for all k, $(m(k))_{k=1}^K \subset \mathbb{N}$, $(H_k)_{k=1}^K$ is a sequence of finite subsets of $\mathbb{N}$ such that $|H_k| \geq M_{m(k)}$ for all k, $H_k \cap H_{k'} = \emptyset$ if $m(k) = m(k'), k \neq k'$, and for all $i \in H_k$*

$$(7.2.1) \qquad |x_{m(k),i}^* \otimes y_j^*((P_{N_k} - P_{N_{k-1}'}) x_{m(k),i} \otimes y_j)| \geq \rho$$

then $K < \max\{2\delta^{-1}, (\|T\|^2 \|T^{-1}\| \|P\| 8/(\delta^{1/2}(4\rho - \rho_1)))^{2p/(p-2)}\}$.

Proof. Fix $m \in \mathbb{N}$. For $\epsilon_i = \epsilon = \rho_1 w_m^{4p/(p-2)}/16$, $C = 1$, $D = \|T\| \|T^{-1}\| \|P\|$, $w_0 = w_{m,j} = w_m$, $r = p$ and $K = \delta(w_m^{2p/(p-2)})^{-1}$, we obtain an integer M_m from Lemma 6.2.

We want to construct blocks as in Lemma 7.1. We are already given the integers N_k and sets H_k so we only need to refine the H_k's to get the sets F_k, define the z_k's as in Lemma 7.1 and check the hypotheses of Lemma 7.1.

First we apply Lemma 6.2. We use the operator $QT^{-1}P(P_{N_{k+1}} - P_{N'_k})T$ and the sequence $(x_{m(k+1),n} \otimes y_j)_{n \in H_{k+1}}$ to obtain a subset F_{k+1} of H_{k+1} of cardinality $K_{k+1} = \delta(w_{m(k+1)}^{2p/(p-2)})^{-1}$ such that

$$(7.2.2) \quad |x^*_{m(k+1),i} \otimes y^*_j(T^{-1}P(P_{N_{k+1}} - P_{N'_k})Tx_{m(k+1),i'} \otimes y_j)|$$
$$< \rho_1 w_{m(k+1)}^{4p/(p-2)+1}/16,$$

for all $i \neq i', i, i' \in F_{k+1}$. (Here the parameters w_{n_i} which occur in Lemma 6.2 are not dependent on the index i.)

This completes the inductive definition of the sets F_k. It remains to verify that our choice of F_{k+1} verifies the hypothesis of Lemma 7.1.

First we have that for $n \in F_{k+1}$, $n \in H_{k+1}$ and thus by 7.2.1

$$x^*_{m(k+1),n} \otimes y^*_j(T^{-1}P(P_{N_{k+1}} - P_{N'_k})Tx_{m(k+1),n} \otimes y_j) \geq \rho$$

By 7.2.2

$$|x^*_{m(k+1),n} \otimes y^*_j(T^{-1}P((P_{N_{k+1}} - P_{N'_k}) \sum_{\substack{s \neq n \\ s \in F_{k+1}}} w_{m(k+1)}^{2/(p-2)}T(x_{m(k+1),s} \otimes y_j)|$$

$$\leq |F_{k+1}|\rho_1 w_{m(k+1)}^{4p/(p-2)+1}/16$$
$$\leq \rho_1 w_{m(k+1)}^{2p/(p-2)}/16$$
$$(7.2.3)$$
$$< \rho_1 w_{m(k+1)}^{2/(p-2)}/4.$$

This shows that the hypotheses of Lemma 7.1 are satisfied with ρ, $\rho' = \rho - \rho_1$, and δ. Thus $K \leq \max\{2\delta^{-1}, (\|T\|^2\|T^{-1}\|\|P\|8/(\delta^{1/2}(4\rho - \rho_1))^{2p/(p-2)}\}$. $\square$

In order to show that there is no isomorphism from $X_p \otimes X_p$ into R_p^α with complemented range, we will show that any isomorphism from $X_p \otimes X_p$ into a $(p,2)$−sum actually must have a large part in finitely many of the summands. The previous lemmas give us tools to use in quantitative gliding hump arguments. Lemma 7.3 is the first in a series of lemmas which estimate how much is mapped outside a finite number of summands.

LEMMA 7.3. *Suppose that T, P, and P_N are as in Lemma 7.1, $0 < \epsilon < 1$, and $\rho_1 > 0$. Then there exists an integer N_0 and integers M_m, such that if*

$$w_j < (4\epsilon - \rho_1)/(12\|T\|\|T^{-1}\|\|P\|),$$

then for each $m \in \mathbb{N}$, and $N \geq N_0$,

$$|\{t \in \mathbb{N} : |x^*_{m,t} \otimes y^*_j(T^{-1}P(I - P_N)Tx_{m,t} \otimes y_j)| > \epsilon\}| \leq M_m.$$

Proof. For each m the integers M_m are chosen by the criteria established in Lemma 7.2 with $\rho = \epsilon$, $\delta = 1/2$, and ρ_1.

Suppose no N_0 works for these M_m. Then for each N there exist an $m \in \mathbb{N}$, $N' \geq N$, and a set $L_m \subset \mathbb{N}$ of cardinality at least M_m such that

$$(7.3.1) \qquad |x_{m,s}^* \otimes y_j^*(T^{-1}P(I - P_{N'})Tx_{m,s} \otimes y_j)| \geq \epsilon$$

for all $s \in L_m$. We will apply this inductively to construct a sequence of pairs of integers (N_k, N_k') and finite subsets of $\mathbb{N}$, (F_k), as in Lemma 7.2 with $\rho = \epsilon$ and $\delta = 1/2$. Suppose we have found $N_1, N_1', \ldots, N_k$ and $F_1, \ldots, F_k$. By assumption for $N = N_k$ there exists an integer m_{k+1}, $N_k' \geq N_k$ and an infinite set $L_{m_{k+1}}$ as above.

By a simple perturbation argument we may assume that each of the elements $Tx_{m_{k+1},i} \otimes y_j, i \in F_{k+1}$ is nonzero in a finite number of the summands Y_n and thus there is an N_{k+1} such that

$$(7.3.2) \qquad P_{N_{k+1}}Tx_{m_{k+1},i} \otimes y_j = Tx_{m_{k+1},i} \otimes y_j$$

for all $i \in F_{k+1}$. This completes the induction step of the construction.

To complete the proof we need to show that if we continue in this way we can find K blocks as in Lemma 7.2 and thus reach a contradiction for K large enough. However note that we have by 7.3.1 and 7.3.2 for $n \in F_{k+1}$

$$|x_{m_{k+1},n}^* \otimes y_j^*(T^{-1}P(P_{N_{k+1}} - P_{N_k'})Tx_{m_{k+1},n} \otimes y_j)|$$
$$= |x_{m_{k+1},n}^* \otimes y_j^*(T^{-1}P(I - P_{N_k'})Tx_{m_{k+1},n} \otimes y_j)| \geq \epsilon.$$

This shows that with $\delta = 1/2$ the hypotheses of Lemma 7.2 are satisfied. Therefore we have a contradiction for large k and the choice of M_m works for some N_0. $\square$

As we noted before Lemma 7.1 the proof can be simplified and the results strengthened if we assume that we have an unconditional basis in the range space.

LEMMA 7.4. *Suppose that there is a constant D such that for all n, Y_n is a subspace of L_p with a D-unconditional basis and that (x_i) and (y_j) are bases of $X_{p,(w_{n,k})}$. Suppose that T, P_N, and P are as in Lemma 7.1, $0 < \epsilon < 1$, and j is such that $w_j < 4\epsilon^2\delta/(9\|T^{-1}\|^2\|P\|^2\|T\|^2)$. Further assume that there are finite sets $N_{i,j}$ for all $i, j \in \mathbb{N}$ such that*

(1) *$T(x_i \otimes y_j)$ is supported in $N(i,j)$*
(2) *$N(i,j) \cap N(i',j') = \emptyset$ if $i' \neq i$ and $j \neq j'$; $i = i'$, $j \neq j'$ and $\max(o(j), o(j')) > o(i)$; or $i \neq i'$, $j = j'$ and $\max(o(i), o(i')) \geq o(j)$.*

Then there exists an integer N_0 such that for all $N \geq N_0$, $\sum_{n \in F} w_{m,n}^{2p/(p-2)} < \infty$ where $F = \{n : |x_n^* \otimes y_j^*(T^{-1}P(I - P_N)Tx_n \otimes y_j)| \geq \epsilon\}$.

Proof. We use arguments like that in the proofs of Lemmas 7.1, 7.2 and 7.3 but we use unconditionality to avoid the use of Lemma 6.2.

Suppose that no such N_0 exists and fix $\delta, 1 \geq \delta > 1/2$. Then we can find a strictly increasing sequence of integers (N_k) and finite disjoint subsets (F_k) of $\mathbb{N}$ such that

$$1 \geq \sum_{s \in F_k} w_s^{2p/(p-2)} > \delta \text{ for each } k,$$

$$N(s,j) \cap N(n,j) = \emptyset \text{ if } s,n \in \cup_k F_k \text{ and } s \neq n,$$

and for all $n \in F_k$,

$$|x_n^* \otimes y_j^*(T^{-1}P(I - P_{N_k})Tx_n \otimes y_j)| \geq \epsilon$$

and

$$P_{N_{k+1}}x_n \otimes y_j = x_n \otimes y_j.$$

Let $x_n' = (I - P_{N_k})Tx_n \otimes y_j$ for all $n \in F_k$ and all k. $(x_n')_{n \in F_k, k \in \mathbb{N}}$ is an unconditional basic sequence in $(\sum Y_n)_{p,2}$. Define an operator S from $[x_n' : n \in F_k, k \in \mathbb{N}]$ into $[x_n \otimes y_j : n \in F_k, k \in \mathbb{N}]$ by $QT^{-1}P$ where Q is the basis projection onto $[x_n \otimes y_j : n \in F_k, k \in \mathbb{N}]$. By Tong's Theorem the diagonal operator S_D defined by $S_D(\sum_n a_n x_n') = \sum_n a_n(x_n^* \otimes y_j^*(T^{-1}Px_n'))x_n \otimes y_j$ is bounded. As in the proof of Lemma 7.1 choose signs ϵ_n such that $\|\sum_{n \in F_k} \epsilon_n w_n^{2/(p-2)}Tx_n \otimes y_j\|_2 \leq (3/2)\|T\|(\sum_{n \in F_k} w_n^{2p/(p-2)}w_j^2)^{1/2}$.

Let

$$z_k' = \sum_{n \in F_k} \epsilon_n w_n^{2/(p-2)} x_n',$$

$$z_k = \sum_{n \in F_k} \epsilon_n w_n^{2/(p-2)} x_n \otimes y_j,$$

and

$$z_k^* = (\sum_{n \in F_k} w_n^{2p/(p-2)})^{-1} \sum_{n \in F_k} \epsilon_n w_n^{2(p-1)/(p-2)} x_n^* \otimes y_j^*.$$

Let Q' be the usual projection onto $[z_k]$. Then

$$Q'S_D z_k' = z_k^*(S_D z_k')z_k = (\sum_{n \in F_k} w_n^{2p/(p-2)})^{-1}(\sum_{n \in F_k} w_n^{2p/(p-2)}x_n^* \otimes y_j^*(T^{-1}Px_n'))z_k.$$

Notice that $|x_n^* \otimes y_j^*(T^{-1}Px_n')| \geq \epsilon$ for all n. Thus

$$\|S_D\| \max\{(\sum_{i=1}^K \|z_i'\|_p^p)^{1/p}, (\sum_{i=1}^K \|z_i'\|_2^2)^{1/2}\} \geq \|Q'S_D(\sum_{i=1}^K z_i')\|$$

$$\geq \|\sum_{i=1}^K (\epsilon)z_k\|$$

$$\geq \epsilon \max\{(K\delta)^{1/p}, (K\delta)^{1/2}\}.$$

Because

$$\|z_i'\|_2 \leq (3/2)\|T\|(\sum_{n \in F_k} w_n^{2p/(p-2)}w_j^2)^{1/2},$$

$$(\sum_{i=1}^K \|z_i'\|_2^2)^{1/2} \leq (3/2)\|T\|K^{1/2}w_j^{1/2}.$$

For $K > \delta^{-1}$, if $(3/2)\|T\|K^{1/2}w_j^{1/2} > \|T\|K^{1/p}$ then we have that

$$\|S_D\|C(3/2)\|T\|K^{1/2}w_j^{1/2} \geq \epsilon(K\delta)^{1/2}.$$

But $w_j < 4\epsilon^2\delta/(9\|S_D\|^2\|T\|^2)$, so we must have $(3/2)K^{1/2}w_j^{1/2} \leq K^{1/p}$ and therefore

$$\|S_D\|C\|T\|K^{1/p} \geq \epsilon(K\delta)^{1/2}$$

or equivalently,

$$K \leq (\|S_D\|C\|T\|/(\epsilon\delta^{1/2}))^{2p/(p-2)}.$$

Thus the construction can only be made finitely many times and the claimed N_0 exists. $\square$

After our detour into the case of unconditional basis, we continue enlarging the sets of indices for which basis vectors are mapped into a finite number of summands. We are working toward finding rich subsets for the index sets in each side of the tensor product for which we have a good estimate. The next lemma gives us rows in a quantitative way.

LEMMA 7.5. *Suppose that $(x_{i,h})$ and $(y_{j,l})$ are standard bases of $X_{p,(w_{n,k})}$. Let T be an isomorphism from $X_{p,(w_{n,k})} \otimes X_{p,(w_{n,k})}$ into $(\sum Y_j)_{p,2,(w_n)}$ satisfying (T2) and let P be a projection onto the range of T. Let P_N denote the projection onto the first N summands of $(\sum Y_n)_{p,2}$. Let $\epsilon > 0$ and let M_j and M_i be the integers given by Lemma 7.2 for $\rho = 3\epsilon/4$, $\delta = 1/2$, and $\rho_1 = \epsilon/4$. Then for each i, j such that $\max\{w_i, w_j\} < (100\|T\|\|T^{-1}\|\|P\|)^{-1}$, there exist two infinite sets H, L and an integer N_0 such that for all $N > N_0$, if $L' \subset L$ and $|L'| \geq M_j$ then*

$$|\{h : |x_{i,h}^* \otimes y_{j,l}^*(T^{-1}P(I - P_N)Tx_{i,h} \otimes y_{j,l})| \geq \epsilon \quad \forall l \in L'\}| < \infty,$$

and if $H' \subset H$ and $|H'| \geq M_i$ then

$$|\{l : |x_{i,h}^* \otimes y_{j,l}^*(T^{-1}P(I - P_N)Tx_{i,h} \otimes y_{j,l})| \geq \epsilon \quad \forall h \in H'\}| < \infty.$$

Moreover, for all $h \in H, l \in L$,

$$|x_{i,h}^* \otimes y_{j,l}^*(T^{-1}P(I - P_{N_0})Tx_{i,h} \otimes y_{j,l})| < \epsilon.$$

Proof. The proof is by induction. Just to get the induction started let $h_1 = 1$. By Lemma 7.3 there is an $N_1 \in \mathbb{N}$ such that for any $N \geq N_1$, $|x_{i,1}^* \otimes y_{j,l}^*(T^{-1}P(I - P_N)Tx_{i,1} \otimes y_{j,l})| < \epsilon$ for all but M_j many l. For $N = N_1$ there are only finitely many l such that the inequality above fails, so let L_1 be the set of l such that $|x_{i,1}^* \otimes y_{j,l}^*(T^{-1}P(I - P_{N_1})Tx_{i,h} \otimes y_{j,l})| < \epsilon$.

Consider the following (non-mutually exclusive) possibilities.

(1) For every infinite $H \subset \mathbb{N}$ and $N \geq N_1$ there are infinitely many $l \in L_1$ and infinite subsets H_l of H such that for all $h \in H_l$,

$$|x_{i,h}^* \otimes y_{j,l}^*(T^{-1}P(I - P_N)Tx_{i,h} \otimes y_{j,l})| < \epsilon.$$

or

(2) There is an $N \geq N_1$, an infinite set H and there are M_j integers l such that

$$|x_{i,h}^* \otimes y_{j,l}^*(T^{-1}P(I - P_N)Tx_{i,h} \otimes y_{j,l})| \geq \epsilon$$

for all $h \in H$.

If the second possibility occurs, we can find $N_1' \geq N_1$, a subset L_1' of L_1 with cardinality M_j, and an infinite subset H_1 of $\mathbb{N}$ such that for all $h \in H_1, l \in L_1'$ $|x_{i,h}^* \otimes y_{j,l}^*(T^{-1}P(I - P_{N_1'})Tx_{i,h} \otimes y_{j,l})| \geq \epsilon$. By applying Lemma 7.3 at most M_j times we can find an integer $N_2 > N_1'$ such that for any $N \geq N_2$ and $l \in L_1'$,

$$|\{h : |x_{i,h}^* \otimes y_{j,l}^*(T^{-1}P(I - P_N)Tx_{i,h} \otimes y_{j,l})| \geq \epsilon/4\}| < M_i'.$$

(We are assuming that the integer M_i' obtained from Lemma 7.3 is not necessarily the same as that we have obtained from Lemma 7.2.) By eliminating a finite number of $h \in H_1$ we may assume that

$$|x_{i,h}^* \otimes y_{j,l}^*(T^{-1}P(I - P_{N_2})Tx_{i,h} \otimes y_{j,l})| < \epsilon/4$$

for all $h \in H_1 \cup \{h_1\}$. Let l_1 be any element of L_1'. Notice that for any $h \in H_1$, we have

$$|x_{i,h}^* \otimes y_{j,l}^*(T^{-1}P(P_{N_2} - P_{N_1'})Tx_{i,h} \otimes y_{j,l})| \geq \epsilon - \epsilon/4 = 3\epsilon/4$$

and thus we have a block as in Lemma 7.2.

If the second possibility does not occur, then we choose $l_1 \in L_1$ and an infinite set H_1 such that for all $h \in H_1$, $|x_{i,h}^* \otimes y_{j,l_1}^*(T^{-1}P(I - P_{N_1})Tx_{i,h} \otimes y_{j,l_1})| < \epsilon$. Let $N_2 = N_1$.

Now we choose h_2, by interchanging the roles of l and h in the argument above.

If there is an $N \geq N_2$, an infinite set $L \subset L_1$ and there are M_i integers h such that $|x_{i,h}^* \otimes y_{j,l}^*(T^{-1}P(I - P_N)Tx_{i,h} \otimes y_{j,l})| \geq \epsilon$ for all $l \in L$, then let $N_2' > N_2$, $H_2' \subset H_1$ with cardinality M_i, and let L_2 be an infinite subset of $L_1 \setminus L_1'$ such that for all $l \in L_2$ and $h \in H_2'$, $|x_{i,h}^* \otimes y_{j,l}^*(T^{-1}P(I - P_{N_1'})Tx_{i,h} \otimes y_{j,l})| \geq \epsilon$. By applying Lemma 7.3 at most M_i times we can find an integer $N_3 > N_2'$ such that for any $N \geq N_3$ and $h \in H_2'$,

$$|\{l : |x_{i,h}^* \otimes y_{j,l}^*(T^{-1}P(I - P_N)Tx_{i,h} \otimes y_{j,l})| \geq \epsilon/4\}| < M_j'.$$

(M_j' is the integer from Lemma 7.3.) By eliminating a finite number of $l \in L_2$ we may assume that $|x_{i,h}^* \otimes y_{j,l}^*(T^{-1}P(I - P_{N_2})Tx_{i,h} \otimes y_{j,l})| < \epsilon/4$ for all $l \in L_2 \cup \{l_1\}$. Let h_2 be any element of H_2'. Notice that for any $l \in L_2$, we have

$$|x_{i,h}^* \otimes y_{j,l}^*(T^{-1}P(P_{N_3} - P_{N_2'})Tx_{i,h} \otimes y_{j,l})| \geq \epsilon - \epsilon/4 = 3\epsilon/4$$

and thus we again have a block as in Lemma 7.2.

Otherwise choose h_2 and and an infinite set $L_2 \subset L_1$ such that for all $l \in L_2, |x_{i,h_2}^* \otimes y_{j,l}^*(T^{-1}P(I - P_{N_2})Tx_{i,h_2} \otimes y_{j,l})| < \epsilon$. Let $N_3 = N_2$.

The remainder of the proof proceeds by alternately choosing elements h_k and l_k as above. Notice that the number of integers k for which we find the set H'_k of cardinality M_i, L'_k of cardinality M_j, respectively, is limited by Lemma 7.2. Therefore, after some finite number of steps, we have that there is an N_{k_0}, and infinite sets L_{k_0} and H_{k_0} such that for any $N \geq N_{k_0}$ and infinite subsets L of L_{k_0} and H of H_{k_0},

$$|\{l \in L_{k_0} : |x_{i,h}^* \otimes y_{j,l}^*(T^{-1}P(I - P_N)Tx_{i,h} \otimes y_{j,l})| \geq \epsilon \text{ for all } h \in H\}| < M_j$$

and

$$|\{h \in H_{k_0} : |x_{i,h}^* \otimes y_{j,l}^*(T^{-1}P(I - P_N)Tx_{i,h} \otimes y_{j,l})| \geq \epsilon \text{ for all } l \in L\}| < M_i.$$

The sets $H = \{h_k : k \geq k_0\}$ and $L = \{l_k : k \geq k_0\}$ and $N_0 = N_{k_0}$ satisfy the conclusion of the lemma. $\square$

The previous lemma gave us an estimate for a row in each factor. Now we move to a rich set in one factor.

LEMMA 7.6. Let $(x_{i,h})$ and $(y_{j,l})$ be equivalent to the standard basis of $X_{p,2,(w_{n,k})}$. Suppose that T, P_N, and P are as in Lemma 7.1, $0 < \epsilon < 1$, and j is such that $w_j = w_{j,l} < (1 - \epsilon)/(3\|S_D\|\|T\|)$. Let (M_i) be the sequence of integers given by Lemma 7.2 for $\rho = 3\epsilon/4, \delta = 1/2$, and $\rho_1 = \epsilon/4$. Then there exists an integer N_0, an infinite subset L of $\mathbb{N}$, and a rich subset K of $\mathbb{N} \times \mathbb{N}$, such that for all $N \geq N_0$, $j \in \mathbb{N}$, there is a set $L' \subset L$, $|L'| < M_j$ and for all $l \notin L'$, $\{h : (i, h) \subset K, |x_{i,h}^* \otimes y_{j,l}^*(T^{-1}P(I - P_N)Tx_{i,h} \otimes y_{j,l})| < \epsilon\}$ is infinite. Moreover, for all $l \in L, (i, h) \in K, |x_{i,h}^* \otimes y_{j,l}^*(T^{-1}P(I - P_{N_0})Tx_{i,h} \otimes y_{j,l})| < \epsilon$.

Proof. By discarding the first few rows of $\mathbb{N} \times \mathbb{N}$, i.e., $\{1, 2, .., k\} \times \mathbb{N}$ for some k, and renumbering we may assume that $w_i = w_{i,h} < (1 - \epsilon)/(3\|T^{-1}\|\|T\|)$ for all $i, h \in \mathbb{N}$. We will inductively construct L and K by using Lemma 7.5 and Lemma 7.2.

First by Lemma 7.5 there exist an integer N_1, an infinite subset L_1 of $\mathbb{N}$ and an infinite subset H_1 of $\{1\} \times \mathbb{N}$ such that for all $N \geq N_1$,

$$\{(h, l) : |x_{1,h}^* \otimes y_{j,l}^*(T^{-1}P(I - P_N)Tx_{i,h} \otimes y_{j,l})| < \epsilon\}$$

does not contain a rectangle $A \times B$ with A infinite and cardinality of B greater than or equal to M_j or with B infinite and cardinality of A greater than or equal to M_1. Also $|x_{1,h}^* \otimes y_{j,l}^*(T^{-1}P(I - P_{N_1})Tx_{i,h} \otimes y_{j,l})| < \epsilon$ for all $l \in L_1$ and $h \in H_1$.

Consider

$$L^N = \{l \in L_1 : \exists K \text{ rich such that}$$
$$|x_{i,h}^* \otimes y_{j,l}^*(T^{-1}P(I - P_N)Tx_{i,h} \otimes y_{j,l})| < \epsilon \qquad \forall (i, h) \in K\}.$$

If for some $N_1' \geq N_1$, $|L_1 \setminus L^{N_1'}| \geq M_j$ then let $L_1' \subset L_1 \setminus L^{N_1'}$ with cardinality M_j. If we enumerate the elements of L_1' as $(l_i)_{i=1}^{M_j}$ we can produce a rich set which is bad for all $l_i \in L_1'$ inductively as follows.

Because $l_1 \notin L^{N_1'}$, there exists a rich set K_1 such that

$$|x_{i,h}^* \otimes y_{j,l_1}^* (T^{-1}P(I - P_{N_1'})Tx_{i,h} \otimes y_{j,l_1})| < \epsilon \qquad \forall (i,h) \in K_1.$$

Because K_1 is a rich set and $l_2 \notin L^{N_1'}$, there is a rich subset K_2 of K_1 such that

$$|x_{i,h}^* \otimes y_{j,l_2}^* (T^{-1}P(I - P_{N_1'})Tx_{i,h} \otimes y_{j,l_2})| < \epsilon \qquad \forall (i,h) \in K_2.$$

Similarly there exists a rich subset K_3 of K_2 such that

$$|x_{i,h}^* \otimes y_{j,l_3}^* (T^{-1}P(I - P_{N_1'})Tx_{i,h} \otimes y_{j,l_3})| < \epsilon \qquad \forall (i,h) \in K_3.$$

Continuing in this way we find a decreasing sequence of rich sets $(K_i)_{i=1}^{M_j}$ such that

$$|x_{i,h}^* \otimes y_{j,l_k}^* (T^{-1}P(I - P_{N_1'})Tx_{i,h} \otimes y_{j,l_k})| < \epsilon \qquad \forall (i,h) \in K_s, s \leq k.$$

Let $K_1' = K_{M_j}$. By at most M_j applications of Lemma 7.3 there exists an integer $N_1'' > N_1'$ such that for all $N \geq N_1''$, $l \in L_1'$, and $i \in K^1$,

$$|\{h : |x_{i,h}^* \otimes y_{j,l}^* (T^{-1}P(I - P_N)Tx_{i,h} \otimes y_{j,l})| \geq \epsilon/4\}| < M_i',$$

where M_i' is given by Lemma 7.3. Let i_2 be the smallest index in $K_1'^1$. By Lemma 7.5 there is an integer $N_2 > N_1''$, an infinite subset L_2 of $L_1 \setminus L_1'$ and an infinite subset H_2 of $\{h : (i_2, h) \in K_1'\}$, such that for any $N \geq N_2$, $\{(h,l) : |x_{i,h}^* \otimes y_{j,l}^* (T^{-1}P(I - P_N)Tx_{i,h} \otimes y_{j,l})| < \epsilon\}$ does not contain a rectangle $A \times B$ with A infinite and cardinality of B greater than or equal to M_j or with B infinite and cardinality of A greater than or equal to M_{i_2}. Also $|x_{i,h}^* \otimes y_{j,l}^* (T^{-1}P(I - P_{N_2})Tx_{i,h} \otimes y_{j,l})| < \epsilon$ for all $l \in L_2$ and $h \in H_2$. By discarding at most a finite number of elements from each row of K_1' we may assume that for all $(i,h) \in K_1'$ and $l \in L_1'$ and for all $(1,h)$ with $h \in H_1$ and $l \in L_2 \cup L_1'$,

$$|x_{i,h}^* \otimes y_{j,l}^* (T^{-1}P(I - P_N)Tx_{i,h} \otimes y_{j,l})| < \epsilon.$$

Observe that for any $(i,h) \in K_1$ we have that for each $l \in L_1'$,

$$|x_{i,h}^* \otimes y_{j,l}^* (T^{-1}P(P_{N_2} - P_{N_1})Tx_{i,h} \otimes y_{j,l})| > \epsilon - \epsilon/4 = 3\epsilon/4.$$

Thus we have a block with respect to l as in Lemma 7.2.

If we cannot find the set L_1' of cardinality M_j as above, choose $l_1 \in L_1$ and let K_1 be a rich subset of L^{N_1} such that $|x_{i,h}^* \otimes y_{j,l_1}^* (T^{-1}P(I - P_{N_1})Tx_{i,h} \otimes y_{j,l_1})| < \epsilon$ for all $(i,h) \in K_1$. Note that we may assume that K_1 is maximal and thus that $(1,h) \in K_1$ for all $h \in H_1$. Let $L_1' = \{l_1\}$.

For each $i \in K_1^1$, $N \geq N_1$, and infinite set $L \subset L_2$ consider the set

$$H(i,N,L) = \{h : |x_{i,h}^* \otimes y_{j,l}^* (T^{-1}P(I - P_N)Tx_{i,h} \otimes y_{j,l})| \geq \epsilon \qquad \forall l \in L\}.$$

If the cardinality of $H(i, N, L)$ is at least M_i for some $i = i_2$, $L = L_1'$ and $N = N_1' \geq N_1$, let H_2' be a subset of $H(i_2, N_1', L_1')$ with cardinality M_i. By Lemma 7.5 there is an integer $N_2 > N_1'$, an infinite subset L_2 of L_1' and an infinite subset H_2 of $\{h : (i_2, h) \in K_1'\}$, such that for any $N \geq N_2$,

$$\{(h, l) : |x_{i,h}^* \otimes y_{j,l}^* (T^{-1} P(I - P_N) T x_{i,h} \otimes y_{j,l})| < \epsilon/2\}$$

does not contain a rectangle $A \times B$ with A infinite and cardinality of B greater than or equal to M_j or with B infinite and cardinality of A greater than or equal to M_{i_2}. Also $|x_{i,h}^* \otimes y_{j,l}^* (T^{-1} P(I - P_{N_2}) T x_{i,h} \otimes y_{j,l})| < \epsilon/2$ for all $l \in L_2$ and $h \in H_2$. By passing to infinite subsets of H_1 and L_2 we may assume that $|x_{i,h}^* \otimes y_{j,l}^* (T^{-1} P(I - P_{N_2}) T x_{i,h} \otimes y_{j,l})| < \epsilon$ for all $l \in L_2 \cup \{l_1\}$ and $h \in H_1$.

Observe that for any $h \in H_2'$ we have that for each $l \in L_2$,

$$|x_{i_2,h}^* \otimes y_{j,l}^* (T^{-1} P(P_{N_2} - P_{N_1}) T x_{i_2,h} \otimes y_{j,l})| > \epsilon - \epsilon/4 = 3\epsilon/4.$$

Thus we have a block with respect to h as in Lemma 7.2.

If neither L_1' of cardinality M_j as in the first case nor H_2' of cardinality M_i as in the second case can be found, then choose an infinite subset L_2 of L_1 and an infinite subset H_2 of $\{h : (i, h) \in K_1\}$ such that $|x_{i,h}^* \otimes y_{j,l}^* (T^{-1} P(I - P_{N_1}) T x_{i,h} \otimes y_{j,l})| < \epsilon$ for all $h \in H_2$ and $l \in L_2$. Let $N_2 = N_1$. let l_1 be any element of L_2 and let K_2 be a maximal rich subset of $\mathbb{N} \times \mathbb{N}$ such that

$$|x_{i,h}^* \otimes y_{j,l}^* (T^{-1} P(I - P_{N_2}) T x_{i,h} \otimes y_{j,l})| < \epsilon$$

for all $(i, h) \in K_2$.

This completes the first full step of our construction. Notice that in each case we have produced an integer N_2, infinite sets L_2, H_1, H_2 of $\mathbb{N}$, a set L_1' containing at least one element and a rich set K_2 such that

$$|x_{i,h}^* \otimes y_{j,l}^* (T^{-1} P(I - P_{N_2}) T x_{i,h} \otimes y_{j,l})| < \epsilon$$

for all $l \in L_2 \cup \{l_1\}$ and $(i, h) \in \{1\} \times H_1 \cup i_2 \times H_2$ and for all $(i, h) \in K_2$ and $l \in L_1'$,

$$|x_{i,h}^* \otimes y_{j,l}^* (T^{-1} P(I - P_{N_2}) T x_{i,h} \otimes y_{j,l})| < \epsilon.$$

We will present one more step of the induction.
Consider

$$L^N = \{l \in L_2 : \exists K \text{ rich such that}$$
$$|x_{i,h}^* \otimes y_{j,l}^* (T^{-1} P(I - P_N) T x_{i,h} \otimes y_{j,l})| < \epsilon \quad \forall (i, h) \in K\}.$$

If for some $N_2' \geq N_2$, $|L_2 \setminus L^{N_2'}| \geq M_j$ then let $L_2' \subset L_2^{N_2'} \setminus L$ with cardinality M_j. If we enumerate the elements of L_2' as $(l_i)_{i=M_j+1}^{2M_j}$ we can produce a rich set K_2' which is bad for all $l_i, M_j < i \leq 2M_j$ inductively as above. By at most M_j applications of Lemma 7.3 there exists an integer $N_2'' > N_2'$ such that for all $N \geq N_2''$, $l \in L_2'$, and $i \in (K_2')^1$,

$$|\{h : |x_{i,h}^* \otimes y_{j,l}^* (T^{-1} P(I - P_N) T x_{i,h} \otimes y_{j,l})| \geq \epsilon/4\}| < M_i'.$$

Let i_3 be the smallest index in $(K_2')^1$. By Lemma 7.5 there is an integer $N_3 > N_2''$, an infinite subset L_3 of $L_2 \setminus L_2'$ and an infinite subset H_3 of $\{h : (i_3, h) \in K_2'\}$, such that for any $N \geq N_3$, $\{(h, l) : |x_{i,h}^* \otimes y_{j,l}^*(T^{-1}P(I - P_N)Tx_{i,h} \otimes y_{j,l})| < \epsilon\}$ does not contain a rectangle $A \times B$ with A infinite and cardinality of B greater than or equal to M_j or with B infinite and cardinality of A greater than or equal to M_{i_3}. Also $|x_{i,h}^* \otimes y_{j,l}^*(T^{-1}P(I - P_{N_3})Tx_{i,h} \otimes y_{j,l})| < \epsilon$ for all $l \in L_3$ and $h \in H_3$. By discarding at most a finite number of elements from each row of K_2' we may assume that for all $(i, h) \in K_2'$ and $l \in L_2'$ and for all (i, h) with $h \in H_i$ for $i = 1, 2$ and $l \in L_3 \cup L_2' \cup L_1'$,

$$|x_{i,h}^* \otimes y_{j,l}^*(T^{-1}P(I - P_N)Tx_{i,h} \otimes y_{j,l})| < \epsilon.$$

Observe that for any $(i, h) \in K_2$ we have that for each $l \in L_2'$,

$$|x_{i,h}^* \otimes y_{j,l}^*(T^{-1}P(P_{N_2} - P_{N_1})Tx_{i,h} \otimes y_{j,l})| > \epsilon - \epsilon/4 = 3\epsilon/4.$$

Thus we have a new block with respect to l as in Lemma 7.2. (If $N_2 = N_1$ this case cannot occur.)

If we cannot find the set L_2' of cardinality M_j as above, choose $l_2 \in L_2$ and let K_2 be a rich subset of L^{N_2} such that $|x_{i,h}^* \otimes y_{j,l_2}^*(T^{-1}P(I - P_{N_1})Tx_{i,h} \otimes y_{j,l_2})| < \epsilon$ for all $(i, h) \in K_2$. Note that we may assume that K_2 is maximal and thus that $(i_s, h) \in K_2$ for all $h \in H_s, s = 1, 2$. Let $L_2' = \{l_2\}$.

For each $i \in (K_2)^1$, $N \geq N_2$, and infinite set $L \subset L_2$ consider the set

$$H(i, N, L) = \{h : |x_{i,h}^* \otimes y_{j,l}^*(T^{-1}P(I - P_N)Tx_{i,h} \otimes y_{j,l})| \geq \epsilon \qquad \forall l \in L\}.$$

If the cardinality of $H(i, N, L)$ is at least M_i for some $i = i_3$, $L = L_2'$ and $N = N_2' \geq N_2$, let H_3' be a subset of $H(i_3, N_2', L_2')$ with cardinality M_i. By Lemma 7.5 there is an integer $N_3 > N_2'$, an infinite subset L_3 of L_2' and an infinite subset H_3 of $\{h : (i_3, h) \in K_2'\}$, such that for any $N \geq N_2$,

$$\{(h, l) : |x_{i,h}^* \otimes y_{j,l}^*(T^{-1}P(I - P_N)Tx_{i,h} \otimes y_{j,l})| < \epsilon/4\}$$

does not contain a rectangle $A \times B$ with A infinite and cardinality of B greater than or equal to M_j or with B infinite and cardinality of A greater than or equal to M_{i_3}. Also $|x_{i,h}^* \otimes y_{j,l}^*(T^{-1}P(I - P_{N_3})Tx_{i,h} \otimes y_{j,l})| < \epsilon/4$ for all $l \in L_3$ and $h \in H_3$. By passing to infinite subsets of H_3 and L_3 we may assume that $|x_{i,h}^* \otimes y_{j,l}^*(T^{-1}P(I - P_{N_3})Tx_{i,h} \otimes y_{j,l})| < \epsilon$ for all $l \in L_3 \cup L_1' \cup L_2'$ and $h \in H_1 \cup H_2$.

Observe that for any $h \in H_3'$ we have that for each $l \in L_3$,

$$|x_{i_3,h}^* \otimes y_{j,l}^*(T^{-1}P(P_{N_2} - P_{N_1})Tx_{i_3,h} \otimes y_{j,l})| > \epsilon - \epsilon/4 = 3\epsilon/4.$$

Thus we have a new block with respect to h as in Lemma 7.2.

If neither L_2' of cardinality M_j as in the first case nor H_3' of cardinality M_i as in the second case can be found, then choose an infinite subset L_3 of L_2 and an infinite subset H_3 of $\{h : (i, h) \in K_2\}$ such that $|x_{i,h}^* \otimes y_{j,l}^*(T^{-1}P(I - P_{N_2})Tx_{i,h} \otimes$

$y_{j,l})| < \epsilon$ for all $h \in H_3$ and $l \in L_3$. Let $N_3 = N_2$. let l_2 be any element of L_3 and let K_3 be a maximal rich subset of $\mathbb{N} \times \mathbb{N}$ such that

$$|x_{i,h}^* \otimes y_{j,l}^* (T^{-1}P(I - P_{N_3})Tx_{i,h} \otimes y_{j,l})| < \epsilon$$

for all $(i, h) \in K_3$.

We have now completed a second full step of the induction. Continuing in this way we produce an increasing sequence of integers (N_k), (i_k) and sequences of sets of integers (L'_k), (H_k), (L_k), and rich sets (K_k). The sequence (N_k) must be eventually constant since each increase in N_k is produced when a new set L'_k of cardinality M_j is found or a new set H'_k of cardinality M_{i_k} is found. If $N_{k_1} < \cdots < N_{k_s}$ and for each $r, 1 \le r \le s$, we have disjoint sets L'_{k_r}, then any $(i, h) \in K_{s+1}$ will give us s blocks as in Lemma 7.2. But Lemma 7.2 gives a bound on the number of such blocks and thus this cannot be the cause of the increase in N_k. Similarly, if we have disjoint sets H'_{k_r} of cardinality M_{i_r} then choosing any $l \in L_{s+1}$ will give us s blocks as in Lemma 7.2. Therefore there is an integer k_0 such that $N_k = N_{k_0}$ for all $k \ge k_0$. Consequently, we can let $N_0 = N_{k_0}$, $L = \{l_r : r \ge k_0\}$ and $K = \cup_{r \ge k_0} \{i_r\} \times H_r$. $\square$

CHAPTER 8

$X_p \otimes X_p$ **IS NOT IN THE SCALE** $R_p^\alpha, \alpha < \omega_1$

In this section we prove our main result and answer a question posed in [BRS]. Before beginning we need a few combinatorial lemmas.

LEMMA 8.1. *If $G \subset \mathbb{N} \times \mathbb{N}$ and there exists an integers M such that if $A \subset \mathbb{N}$ with cardinality M and $B \subset \mathbb{N}$ which is infinite, then $A \times B \cap G \neq \emptyset$ and $B \times A \cap G \neq \emptyset$. Then there exist infinite subsets of $\mathbb{N}$, H, L such that $H \times L \subset G$.*

Proof. Observe that the hypothesis implies the following.

Given $N \subset \mathbb{N}$ with cardinality greater than $M - 1$ and any infinite subset J of $\mathbb{N}$ then there is an element n of N and an infinite set $J' \subset J$. such that $(n, j) \in G$ for all $j \in J'$.

Indeed, enumerate the elements of N as $n_1, n_2, \ldots n_M$. By hypothesis $\cap_r \{j \in J : (n_r, j) \notin G\}$ is finite and

$$\cap_r \{j \in J : (n_r, j) \notin G\} = J \setminus (\{j \in J : (n_1, j) \in G\} \cup \{j \in J : (n_1, j) \notin G,$$
$$(n_2, j) \in G\} \cup \cdots \cup \{j \in J : (n_r, j) \notin G \text{ for } r = 1, 2, \ldots, M - 1, (n_M, j) \in G\}.$$

thus one of the sets

$$\{j \in J : (n_1, j) \notin G, (n_2, j) \in G\} \cup \ldots \{j \in J : (n_r, j) \notin G$$
$$\text{for } r = 1, 2, \ldots, s, (n_{s+1}, j) \in G\}$$

is infinite.

To construct the sets H, L we alternately select infinite sets H_k and L_k and elements h_k and l_k such that

    (1) (H_k) and (L_k) are decreasing,
    (2) $(h_k, l) \in G$ for all $l \in L_{k+1}$,
    (3) $(h, l_k) \in G$ for all $h \in H_k$,
    (4) $h_k \in H_k$
    (5) $l_k \in L_k$

Begin with $H_0 = \mathbb{N}$ and $L_1 = \mathbb{N}$, and use the principle to find $l_1 \in \{1, 2, \ldots M\}$ and an infinite set H_1 such that $(h, l_1) \in G$ for all $h \in H_1$. Next let H_1' be a subset of H_1 with cardinality M and let $L_0 = \mathbb{N} \setminus \{l_1\}$. By the principle there exists an element h_1 of H_1' and an infinite subset L_2 of L_1 such that $(h_1, l) \in G$ for all $l \in L_2$. Next let L_2' be a subset of $L_2 \setminus \{l_1\}$ with cardinality M. By the principle there is an infinite subset H_2 of $H_1 \setminus \{h_1\}$ and $l_2 \in L_2'$ such that $(h, l_2) \in G$ for all $h \in H_2$. Clearly this procedure will produce the required sequences. $\square$

LEMMA 8.2. *Suppose that $G \subset \mathbb{N} \times \mathbb{N} \times \mathbb{N}$ and $N \in \mathbb{N}$ then one of the following holds*

(1) *There exists a rich subset $K \subset \mathbb{N} \times \mathbb{N}$ such that for each $(j, l) \in K$, there are at least N elements $n \in \mathbb{N}$ such that $(n, j, l) \notin G$.*

(2) *There exist $M \subset \mathbb{N}$, M infinite, and a rich subset $K \subset \mathbb{N} \times \mathbb{N}$ such that for each $n \in M$ and $(j, l) \in K$, $(n, j, l) \in G$.*

Proof. We begin by trying to directly satisfy the first alternative. For each $j \in \mathbb{N}$ let

$$L_j = \{l : \text{ there exists } F \subset \mathbb{N}, |F| \geq N \text{ such that } (n, j, l) \notin G \text{ for all } n \in F\}.$$

Let $J = \{j : |L_j| = \infty\}$. If J is infinite, then we define $K = \{(j, l) : j \in J, l \in L_j\}$. K is clearly rich and satisfies the first criterion.

If J is finite, discard all (n, j, l) such that $j \in J$. Because $\mathbb{N} \setminus L_j$ is infinite for all $j \notin J$, we can without loss of generality assume that $J = \emptyset$ and $L_j = \emptyset$ for all j. We need to remove another set of elements before we try to satisfy the second criterion.

Suppose that there exists some $n_1 \in \mathbb{N}$ such that $J_1 = \{j : |\{l : (n_1, j, l) \notin G\}| = \infty\}$ is infinite. Then we replace $\mathbb{N} \times \mathbb{N} \times \mathbb{N}$ by $\mathbb{N} \setminus \{n_1\} \times \{(j, l) : j \in J_1, (n_1, j, l) \notin G\}$. If there exists $n_2 \in \mathbb{N} \setminus \{n_1\}$ such that $J_2 = \{j : j \in J_1, |\{l : (n_i, j, l) \notin G, i = 1, 2\}| = \infty\}$ we remove n_2 and consider only $j \in J_2$ and l such that $(n_i, j, l) \notin G, i = 1, 2$. Because we have eliminated the pairs of coordinates which appear N or more times, this process must stop in at most $N - 1$ steps. Thus there exist some $k < N$, $N' = \mathbb{N} \setminus \{n_1, \ldots, n_k\}$, an infinite set $J' = J_k \subset \mathbb{N}$, and for each $j \in J'$, $L_j \subset \mathbb{N}$ infinite such that for each $l \in L_j$, $(n, j, l) \notin G$ for at most $N - 1$ elements $n \in N'$ and for each $n \in N'$ there are only finitely many $j \in J'$ such that $\{l : (n, j, l) \notin G\}$ is infinite.

We now want to construct the sets M and K by a procedure like that used to count the rationals. Choose $m_1 \in N'$. Let $J_1' = \{j : |\{l : (m_1, j, l) \notin G\}| < \infty\}$. Then by assumption J_1' contains all but finitely many elements of J'. Let $L_{j,1} = \{l : (m_1, j, l) \in G\}$ for each $j \in J_1'$. Choose $j_1 \in J_1'$ and then we can find an infinite subset L_1 of $L_{j_1,1}$ and a co-finite subset N_1' of N' such that for each $n \in N_1'$ $(n, j_1, l) \notin G$ for at most finitely many $l \in L_{j_1,1}$ as follows. If $L_1 = L_{j_1,1}$ and $N_1' = N'$ work, we are done. If not there is some $k_1 \in N'$ such that $K_1 = \{l : (k_1, j_1, l) \notin G\}$ is infinite. If $N_1' = N' \setminus \{k_1\}$ and $L_1 = K_1$ work we are done. If not there exists, $k_2 \in N_1'$, $k_2 \neq k_1$, such that $K_2 = \{l : (k_i, j_1, l) \notin G, i = 1, 2\}$ is infinite. As in the earlier argument this can continue only at most $N - 1$ times. Now choose $l_{j_1,1} \in L_1$. Let $N_1'' = \{n : n \in N_1', (n, j_1, l_1) \in G\}$. Again we have lost at most finitely many elements of N_1'.

Next choose $m_2 \in N_1''$, let $J_2' = \{j : j \in J_1', |\{l : (m_2, j, l) \notin G\}| < \infty\}$. Then by assumption J_2' contains all but finitely many elements of J_1'. For $j \neq j_1$ let $L_{j,2} = \{l : (m_2, j, l) \in G\}$ for each $j \in J_2'$ and $L_{j_1,2} = \{l : l \in L_1, (m_2, j, l) \in G\}$. Note that $l_{j_1,1} \in L_{j_1,2}$. Next choose $j_2 \in J_2', j_2 \neq j_1$. As above we can find a co-finite subset N_2 of N_1'' and find an infinite subset L_2 of $L_{j_2,2}$ such that for each $n \in N_2'$, $(n, j_2, l) \notin G$ for at most finitely many $l \in L_{j_2,1}$. Now choose $l_{j_1,2} \in L_{j_1,2}, l_{j_1,2} \neq l_{j_1,1}$ and $l_{j_2,1} \in L_2$. Let $N_2'' = N_2' \setminus \{n : (n, j_1, l_{j_1,2}) \notin$

G or $(n, j_2, l_{j_2,1}) \notin G\}$. This removes at most a finite number of elements from N_2' but not m_1 or m_2.

The remainder of the argument consists of inductively choosing as we have done above new m_i's, new j_i's, and corresponding $l_{j_i,k}$, so that in the end $M = \{m_1, m_2, \dots\}$ and $K = \{(j_i, l_{j_i,k}) : i, k \in \mathbb{N}\}$ satisfy the second criterion. We leave the details to the reader. $\square$

We are finally ready to prove our main result.

THEOREM 8.3. *Suppose that there is a constant D such that for all n, Y_n is a subspace of L_p with a D-unconditional basis . If T is an isomorphism from $X_p \otimes X_p$ into $(\sum Y_n)_{p,2,(w_n)}$ and P is a projection onto the range of T, then there exists an integer N and a subspace Z of $X_p \otimes X_p$, isomorphic to $X_p \otimes X_p$ such $P_N T|_Z$ is an isomorphism and $P_N T(Z)$ is complemented.*

Proof. We will use the standard basis of $X_{p,2,(w_{n,k})}$ where as usual $w_{n,k} = w_n$ and (w_n) decreases to 0. Let $(x_{i,h})$ and $(y_{j,l})$ be two copies of the basis. By Proposition 6.3 it is sufficient to find $N \in \mathbb{N}, \epsilon > 0$, and rich subsets of $\mathbb{N} \times \mathbb{N}$, M, K, such that

$$|x_{i,h}^* \otimes y_{j,l}^* (T^{-1} P P_N T x_{i,h} \otimes y_{j,l})| > \epsilon$$

for all $(i, h) \in M, (j, l) \in K$. In this proof we will take $\epsilon = 1/4$. We proceed by induction as in the proof of Proposition 6.3. Lemmas 7.2 and 7.4 will be used to show that a uniform N works. We will use the two player game approach with two interwoven games as we did in the proof of Proposition 6.3. The proof is essentially combinatorial so that topological condition in the definition of the games will be irrelevant. (We could take X_0 to be the weak closure of the X_p basis, $(x_{m,n})$ or $(y_{j,l})$, and use constant functions, but we will not define the functions at all.) Lemmas 7.2 and 7.4 require that we use only small w_k thus we immediately discard all indices $(i, h), (j, l)$ for which w_i or w_j is larger than $(32\|T\|\|T^{-1}\|\|P\|)^{-1}$. This does not affect the isomorphic type of the span of the remaining basis vectors. Therefore we will just assume that the index sets are again $\mathbb{N} \times \mathbb{N}$. Let (M_m) be the sequence of integers determined by Lemma 7.2 for $\rho = 1/2, \rho_1 = 1/4$ and $\delta = 1/2$.

To reduce the notation a little define

$$f(N, i, h, j, l) = |x_{i,h}^* \otimes y_{j,l}^* (T^{-1} P(I - P_N) T x_{i,h} \otimes y_{j,l})|$$

and

$$g(N, i, h, j, l) = |x_{i,h}^* \otimes y_{j,l}^* (T^{-1} P P_N T x_{i,h} \otimes y_{j,l})|$$

for all $i, h, j, l \in \mathbb{N}$.

Let s_1 be chosen according the game argument as in the proof of Proposition 5.4 and let $S_{X,1}'$ be the set in $\mathcal{S}_X$. By Lemma 7.4 there is an integer N_1 and a set $S_Y \in \mathcal{S}_Y$, such that

$$f(N_1, x_{\phi(s)}, y_{\phi(t)}) < 1/4$$

for all $t \in S_Y$ and $s = s_1$.

Let

$$G = \{(h, j, l) : \exists t \in S_Y, s \in S'_{X,1} \text{ such that } \phi(s) = (\phi(s_1)^1, h),$$
$$\phi(t) = (j, l), g(N_1, \phi(s), \phi(t)) \geq 1/4\}.$$

We apply Lemma 8.2 to G with $N = M_{\phi(s_1)^1}$ and with $\mathbb{N} \times S_Y$ in place of $\mathbb{N} \times \mathbb{N} \times \mathbb{N}$.

If alternative (2) occurs, let H_1 be the infinite subset and S'_Y be $\phi^{-1}(K)$, where K is the rich subset. (We assume that K is maximal for H_1 and then that H_1 is maximal for this maximal K.) Let $S''_{X,1} = S'_{X,1} \setminus \{(\phi(s_1)^1, h) : h \notin H_1\}$. Let $N_2 = N_1$. (This is a notational convenience.) Observe that we may assume that $\phi(s_1)^2 \in H_1$.

If alternative (1) occurs and not (2), let K be the rich subset of $\phi(S_Y)$. Then by Lemma 7.6 there is an integer N_2, an infinite subset H_1 of $\{h : \exists s \in S'_{X,1} \ni \phi(s) = (\phi(s_1)^1, h)\}$ and a rich set $K' \subset K$, such that for any $N \geq N_2$, for all j, all but at most $M_{\phi(s_1)}$ indices $h \in H_1$, $f(N, \phi(s_1)^1, h, j, l) < 1/4$ for infinitely many l with $(j, l) \in K'$. Moreover, $f(N_2, \phi(s_1)^1, h, j, l) < 1/4$ for all $i \in H_1$, $(j, l) \in K'$. Let $S'_Y = \phi^{-1}(K')$ and $S''_{X,1} = S'_{X,1} \setminus \{(\phi(s_1)^1, h) : h \notin H_1\}$.

Observe that in both cases we have that for the integer N_2, $g(N_2, x_{\phi(s)}, y_{\phi(t)}) \geq 1/4$ for all $t \in S'_Y$, and all $s \in S''_{X,1}$ such that $\phi(s)^1 = \phi(s_1)^1$. This means that we can allow the second player in X-game to choose any new element s with $\phi(s)^1 = \phi(s_1)^1$ from $S''_{X,1}$. Also note that if alternative (2) failed, then for any $t \in S'_Y$, there exist a set of $M_{\phi(s_1)}$ indices $H'_1 \subset \mathbb{N}$ such that

$$f(N_2, \psi(s_1)^1, h, \phi(t)) < 1/4$$

and

$$g(N_1, \phi(s_1)^1, h, \phi(t)) < 1/4,$$

for all $h \in H'_1$. Therefore

$$|x^*_{\phi(s_1)^1, h} \otimes y^*_{\phi(t)}(T^{-1}P(P_{N_2} - P_{N_1})Tx_{\phi(s_1)^1, h} \otimes y_{\phi(t)})| > 1/2,$$

for all $h \in H'_1$. Thus we have one block for the X-game as in Lemma 7.2.

Now that we have started the construction, we can make further steps a little more regular.

Let

$$T_0 = \{t : t \in S'_Y \text{ and }$$
$$\phi(\{s : s \in S''_{X,1}, g(N_2, \phi(s), \phi(t)) \geq 1/4\}) \text{ contains a rich set}\}.$$

If $\phi(T_0)$ contains a rich set, let K be a maximal one and let $S_{Y,1} = \phi^{-1}(K)$. If not, then $\phi(S'_Y \setminus T_0)$ contains a rich set, K. Let $S_{Y,1} = \phi^{-1}(K)$.

Player 2 in the Y-game chooses $t_1 \in S_{Y,1}$ and $S'_{Y,1} \in \mathcal{S}_Y$ such that $S'_{Y,1} \subset S_{Y,1}$.

Let

$$G = \{(l, i, h) : \exists t \in S'_{Y,1}, s \in S''_{X,1}, \phi(t) = (\phi(t_1)^1, l),$$
$$\phi(s) = (i, h), g(N_2, \phi(s), \phi(t)) \geq 1/4\}.$$

We apply Lemma 8.2 to G with $N = M_{\phi(t_1)^1}$ and with $\mathbb{N} \times S''_{X,1}$ in place of $\mathbb{N} \times \mathbb{N} \times \mathbb{N}$.

If alternative (2) occurs, let L_1 be the infinite subset and

$$S'''_{X,1} = \{s \in S''_{X,1} : \phi(s)^1 = \phi(s_1)^1\} \cup \phi^{-1}(K),$$

where K is the (maximal) rich subset. (We know that if $s \in S''_{X,1}$ and $\phi(s)^1 = \phi(s_1)^1$, then $g(N_2, \phi(s), \phi(t)) \geq 1/4$ and thus $(l, \phi(s)) \in G$.) Let $S''_{Y,1} = S'_{Y,1} \setminus \{(\phi(t_1)^1, l) : l \notin L_1\}$. Let $N_3 = N_2$. (This is again a notational convenience.) Observe that we may assume that $\phi(t_1)^2 \in L_1$ by making L_1 maximal.

If alternative (1) occurs and not (2), let K be the rich subset of $\phi(S''_{X,1})$. Then by Lemma 7.6 there is an integer $N_3 > N_2$, an infinite subset L_1 of $\mathbb{N}$ and a rich set $K' \subset K$, such that for any $N \geq N_3$, for all $i \in \phi(S''_{X,1})^1$, all but at most $M_{\phi(t_1)}$ indices $l \in L_1$, $f(N, i, h, \phi(t_1)^1, l) < 1/4$ for infinitely many h with $(i, h) \in K'$. Moreover, $f(N_3, i, h, \phi(t_1)^1, l) < 1/4$ for all $l \in L_1$, $(i, h) \in K'$. Let

$$S'''_{X,1} = \phi^{-1}(K')$$

and

$$S''_{Y,1} = S'_{Y,1} \setminus \{(\phi(t_1)^1, l) : l \notin L_1\}.$$

Observe that in both cases we have that for the integer N_3, $g(N_3, \phi(s), \phi(t)) \geq 1/4$ for all $s \in S'''_{X,1}$, and all $t \in S''_{Y,1}$ such that $\phi(t)^1 = \phi(t_1)^1$. This means that we can allow the second player in Y-game to choose any new element t with $\phi(t)^1 = \phi(t_1)^1$ from $S''_{Y,1}$. Also note that if alternative (2) failed, then for any $s \in S'''_{X,1}$, there exist a set of $M_{\phi(t_1)}$ indices $L'_1 \subset \mathbb{N}$ such that

$$f(N_3, \phi(s), \phi(t_1)^1, l) < 1/4$$

and

$$g(N_2, \phi(s), \phi(t_1)^1, l) < 1/4,$$

for all $l \in L'_1$. Therefore

$$|x^*_{\phi(s)} \otimes y^*_{\phi(t_1)^1, l}(T^{-1}P(P_{N_3} - P_{N_2})Tx_{\phi(s)} \otimes y_{\phi(t_1)^1, l})| > 1/2,$$

for all $l \in L'_1$. Thus we have one block for the Y-game as in Lemma 7.2.

The idea of the proof is to continue to play the two games as above. Each time we are forced to take alternative (1) in the application of Lemma 8.2, we produce a new block as in Lemma 7.2. However Lemma 7.2 tells us that this cannot go on happening. Thus eventually only alternative (2) occurs in each game and we are free to construct the required basic sequences. In order to play the games according to the rules we will fatten the sets $S_{X,r}, S_{Y,r}$, so that they will contain the previously chosen elements. However in the end we will discard the elements chosen before the integer N_k has been fixed.

We will do a few more steps of the induction in order to cover a few cases which have yet to arise.

Let

$$S_0 = \{s : s \in S'''_{X,1} \text{ and}$$
$$\phi(\{t : t \in S''_{Y,1}, f(N_3, \phi(s), \phi(t)) < 3/4\}) \text{ contains a rich set}\}.$$

If $\phi(S_0)$ contains a rich set, let K be a maximal one and let $S_{X,2} = \{s_1\} \cup \{s : s \in S''_{X,1}, \phi(s)^1 = \phi(s_1)^1\} \cup \phi^{-1}(K)$. If not, then $S'''_{X,1} \setminus S_0$ contains a rich set, K. Let $S_{X,2} = \{s_1\} \cup \{s : s \in S''_{X,1}, \phi(s)^1 = \phi(s_1)^1\} \cup \phi^{-1}(K)$.

Player 2 in the X-game chooses $s_2 \in S_{X,2}$ and $S'_{X,2} \in \mathcal{S}_X$ such that $S'_{X,2} \subset S_{X,2}$.

If it happens that $\phi(s_1)^1 = \phi(s_2)^1$ and $N_3 = N_2$, there is nothing to do. Thus we simply let $S'''_{Y,2} = S''_{Y,1}$. Otherwise, let

$$G = \{(h, j, l) : \exists t \in S''_{Y,1}, s \in S'_{X,2}, \phi(s) = (\phi(s_2)^1, h),$$
$$\phi(t) = (j, l), f(N_3, \phi(s), \phi(t)) < 3/4\}.$$

We apply Lemma 8.2 to G with $N = M_{\phi(s_2)^1}$ and with $(\phi(S'_{X,2}) \cap (\{\phi(s_2)^1\} \times \mathbb{N})) \times S''_{Y,1}$ in place of $\mathbb{N} \times \mathbb{N} \times \mathbb{N}$.

If alternative (2) occurs, let H_2 be the infinite subset and $S'''_{Y,1}$ be $\phi^{-1}(K)$, where K is the maximal rich subset. Let $S''_{X,2} = S'_{X,2} \setminus \{(\phi(s_2)^1, h) : h \notin H_2\}$. Let $N_4 = N_3$. Observe that we may assume that $\phi(s_2)^2 \in H_2$.

If alternative (1) occurs and not (2), let K be the rich subset of $\phi(S''_{Y,1})$. Then by Lemma 7.6 there is an integer $N_4 > N_3$, an infinite subset H_2 of $\mathbb{N}$ and a rich set $K' \subset K$, such that for any $N > N_4$, for all j, all but at most $M_{\phi(s_2)}$ indices $h \in H_1$, $f(N, \phi(s_2)^1, h, j, l) < 1/4$ for infinitely many l with $(j, l) \in K'$. Moreover, $f(N_4, \phi(s_2)^1, h, j, l) < 1/4$ for all $h \in H_2$, $(j, l) \in K'$. Let $S'''_{Y,1} = \phi^{-1}(K')$ and $S''_{X,2} = S'_{X,2} \setminus \{(\phi(s_2)^1, h) : h \notin H_2\}$.

Observe that in all three cases we have that for the integer N_4,

$$g(N_4, x_{\phi(s)}, y_{\phi(t)}) \geq 1/4$$

for all $t \in S'''_{Y,1}$, and all $s \in S''_{X,2}$ such that $\phi(s)^1 = \phi(s_2)^1$. This means that we can allow the second player in X-game to choose any new element s with $\phi(s)^1 = \phi(s_r)^1$, $r = 2$ (or $r = 1, 2$ if $N_4 = N_3$) from $S''_{X,2}$. Also note that if alternative (2) failed, then for any $t \in S'_Y$, there exist a set of $M_{\phi(s_2)}$ indices $H'_2 \subset \mathbb{N}$ such that

$$f(N_4, \phi(s_2)^1, h, \phi(t)) < 1/4$$

and

$$g(N_3, \phi(s_2)^1, h, \phi(t)) < 1/4,$$

for all $h \in H'_2$. Therefore

$$|x^*_{\phi(s_2)^1, h} \otimes y^*_{\phi(t)}(T^{-1}P(P_{N_4} - P_{N_3})Tx_{\phi(s_2)^1, h} \otimes y_{\phi(t)})| > 1/2,$$

for all $h \in H'_2$. Thus for any $t \in S'''_{Y,1} \subset S'_Y$, we have one or two blocks (depending on whether alternative (2) has failed two times in X-game) for the X-game as in Lemma 7.2.

Let

$$T_0 = \{t : t \in S'''_{Y,1} \text{ and }$$

$$\phi(\{s : s \in S''_{X,2}, g(N_4, \phi(s), \phi(t)) \geq 1/4\}) \text{ contains a rich set}\}.$$

If $\phi(T_0)$ contains a rich set, let K be a maximal rich subse and let $S_{Y,2} = \phi^{-1}(K) \cup \{t_1\} \cup \{t \in S'''_{Y,1} : \phi(t)^1 = \phi(t_1)^1\}$. If not, then $\phi(S'''_{Y,1} \setminus T_0)$ contains a rich set, K. Let $S_{Y,2} = \phi^{-1}(K) \cup \{t_1\} \cup \{t \in S'''_{Y,1} : \phi(t)^1 = \phi(t_1)^1\}$.

Player 2 in the Y-game chooses $t_2 \in S_{Y,2}$ and $S'_{Y,2} \in \mathcal{S}_Y$ such that $S'_{Y,2} \subset S_{Y,2}$. If $\phi(s_2)^1 = \phi(s_1)^1$ and $N_3 = N_4$, then let $S'''_{X,2} = S''_{X,2}$. otherwise let

$$G = \{(i, h, l) : \exists t \in S'_{Y,2}, s \in S''_{X,2}, \phi(t) = (\phi(t_2)^1, l),$$

$$\phi(s) = (i, h), g(N_4, \phi(s), \phi(t)) \geq 1/4\}.$$

We apply Lemma 8.2 to G with $N = M_{\phi(t_2)^1}$ and with $(\phi(S'_{Y,2}) \cap (\{\phi(t_2)^1\} \times \mathbb{N})) \times S''_{X,2}$ in place of $\mathbb{N} \times \mathbb{N} \times \mathbb{N}$.

If alternative (2) occurs, let L_2 be the infinite subset and $S'''_{X,2}$ be $\phi^{-1}(K)$, where K is the maximal rich subset. Let $S''_{Y,2} = S'_{Y,2} \setminus \{(\phi(t_2)^1, l) : l \notin L_2\}$. Let $N_5 = N_4$. (This is again a notational convenience.) Observe that we may assume that $\phi(t_2)^2 \in L_2$ by making L_2 maximal.

If alternative (1) occurs and not (2), let K be the rich subset of $\phi(S''_{X,2})$. Then by Lemma 7.6 there is an integer N_5, an infinite subset L_2 of $\mathbb{N}$ and a rich set $K' \subset K$, such that for any $N \geq N_5$, for all $i \in \phi(S''_{X,2})^1$, all but at most $M_{\phi(t_2)}$ indices $l \in L_2$, $f(N, i, h, \phi(t_2)^1, l) < 1/4$ for infinitely many h with $(i, h) \in K'$. Moreover, $f(N_5, i, h, \phi(t_2)^1, l) < 1/4$ for all $l \in L_2$, $(i, h) \in K'$. Let $S'''_{X,2} = \phi^{-1}(K')$ and $S''_{Y,2} = S'_{Y,2} \setminus \{(\phi(t_2)^1, l) : l \notin L_2\}$.

Observe that in all three cases we have that for the integer N_5, $g(N_5, \phi(s), \phi(t)) \geq 1/4$ for all $s \in S'''_{X,2}$, and all $t \in S''_{Y,2}$ such that $\phi(t)^1 = \phi(t_2)^1$. This means that we can allow the second player in Y-game to choose any new element t with $\phi(t)^1 = \phi(t_2)^1$ from $S''_{Y,2}$. Also note that if alternative (2) failed, then for any $s \in S'''_{X,2}$, there exist a set of $M_{\phi(t_2)}$ indices $L'_2 \subset \mathbb{N}$ such that

$$f(N_5, \phi(s), \phi(t_2)^1, l) < 1/4$$

and

$$g(N_4, \phi(s), \phi(t_2)^1, l) < 1/4,$$

for all $l \in L'_1$. Therefore

$$|x^*_{\phi(s)} \otimes y^*_{\phi(t_2)^1, l}(T^{-1}P(P_{N_5} - P_{N_4})Tx_{\phi(s)} \otimes y_{\phi(t_2)^1, l})| > 1/2,$$

for all $l \in L'_2$. Thus for any $s \in S'''_{X,2} \subset S'''_{X,1}$, we have one or two blocks (depending on whether alternative (2) has failed two times in Y-game) for the Y-game as in Lemma 7.2.

We have now established the pattern of the induction. As we have noted earlier alternative (2) of Lemma 8.2 can not fail infinitely many times, thus for some

k_0, $N_k = N_{k_0}$ for all $k \geq k_0$. Also $f(N_{k_0}, \phi(s_k), \phi(t_{k'})) < 3/4$ for all $k, k' \geq k_0$. Therefore $g(N_{k_0}, \phi(s_k), \phi(t_{k'})) > 1 - 3/4 = 1/4$ for all $k, k' \geq k_0$. Because the games must yield a sequence in $\mathcal{S}_X$, $\mathcal{S}_Y$, respectively, and removing finitely many rows from such a set is still such a set, $(x_{\phi(s_k)})_{k \in S}$ and $(y_{\phi(t_k)})_{k \in T}$, where $S = \{s_r : \phi(s_r)^1 \neq \phi(s_k)^1 \quad \forall k < k_0\}$ and $T = \{t_r : \phi(t_r)^1 \neq \phi(t_k)^1 \quad \forall k < k_0\}$, are equivalent to bases of X_p and Proposition 6.3 concludes the proof. $\square$

COROLLARY 8.4. *For all* $\alpha < \omega_1$, $X_p \otimes X_p$ *is not isomorphic to a complemented subspace of* R_p^α.

Proof. We know that for each $\alpha < \omega_1$, R_p^α is isomorphic to an $(p, 2)$ sum of spaces $R_p^{\alpha_n}$ where for each n, $R_p^{\alpha_n}$ is not isomorphic to R_p^α. If $X_p \otimes X_p$ were isomorphic to a complemented subspace of some R_p^γ, $\gamma < \omega_1$, let α be the smallest such ordinal. Since R_p^α is isomorphic to $(\sum R_p^{\alpha_n})_{p,2}$, Theorem 8.3 implies that $X_p \otimes X_p$ is isomorphic to a complemented subspace of $(\sum_{n=1}^N R_p^{\alpha_n})_{p,2}$, for some $N \in \mathbb{N}$. However this space is isomorphic to $R_p^{\alpha'}$ for some α' such that $\alpha' + \omega \leq \alpha$. This contradicts the choice of α. $\square$

CHAPTER 9

FINAL REMARKS AND OPEN PROBLEMS

In this paper we have answered some of the questions posed in [BRS], but there are many more questions that are raised by this work.

1. In Section 1 we note that the best projection onto $\otimes^k X_p$ has norm which grows with k. Thus it is natural to ask: Is $\otimes^k X_p$ isomorphic to a well complemented subspace of L_p? More precisely, is there a constant C and subspaces $Y_k, k \in \mathbb{N}$ of L_p such that $\otimes^k X_p$ is isomorphic to Y_k and there is a projection P_k of L_p onto Y_k with $\|P_k\| \leq C$?

2. In [S] Schechtman uses spaces of the form

$$\left(\sum\left(\sum\cdots\left(\sum\ell_{r_1}\right)_{r_2}\cdots\right)_{r_{n-1}}\right)_{r_n}$$

in order to distinguish the spaces $\otimes^k X_q$, $k \in \mathbb{N}$. If $2 \geq r_1 > r_2 > \cdots > r_1 > q$, is $\left(\sum\left(\sum\cdots\left(\sum\ell_{r_1}\right)_{r_2}\cdots\right)_{r_{n-1}}\right)_{r_n}$ isomorphic to a subspace of $R_q^{\omega n}$? For $n = 1$ this is a result of Rosenthal [R2] and it is not hard to see that for $n = 2$ and $r_1 = 2$ or $r_2 = q$ that it is true. However we do not know whether there is any $\alpha < \omega_1$, such that for $2 > r_1 > r_2 > q$, $\left(\sum\ell_{r_1}\right)_{r_2}$ is isomorphic to a subspace of R_q^α. Notice that were it the case that there is no such α, then Corollary 8.4 would follow.

3. The proof of Theorem 8.3 that we have presented uses the unconditional basis assumption very sparingly. We had hoped to eliminate it altogether. Is the assumption that each Y_n have a D-unconditional basis necessary? Can the proof of Theorem 8.3 be simplified substantially if we make more use of the assumption that the spaces Y_n have unconditional bases?

4. In Chapter 5 we introduce a general framework for gliding hump type arguments, but we do not carry the work very far. What are good classes $\mathcal{S}$ for the natural bases of spaces such as $\otimes^k X_p$, $\left(\sum\left(\sum\cdots\left(\sum\ell_{r_1}\right)_{r_2}\cdots\right)_{r_{n-1}}\right)_{r_n}$, spaces modeled on trees, etc.?

5. This paper shows that at least for certain questions the spaces R_p^α are similar enough to X_p that the techniques originated by Rosenthal can be adapted for use with these spaces. J. T. Woo [Wo1], [Wo2], showed that X_p is just one of a collection of modular sequence spaces with similar properties. Many of the arguments in this paper are really about multiple norm spaces. Thus it is likely that much of it would generalize to a class of spaces where p and 2 are replaced by p and r or perhaps by spaces which are defined by families of indices.

6. Suppose that P is a projection on $X_p \otimes X_p$. Is there a complemented subspace Z of $X_p \otimes X_p$ which is isomorphic to $X_p \otimes X_p$ and is contained in the range

of P or the range of $I - P$? Because of Propostions 6.3 and 6.8, we think that it is very likely that this is true. The main difficulty remaining seems to be a combinatorial problem. Suppose that $G \subset \mathbb{N}^4$ and ϕ is a bijection from $\mathbb{N}$ onto $\mathbb{N} \times \mathbb{N}$ as in Section 5. Are there infinite subsets K, L of $\mathbb{N}$, such that

$$\{(\phi(k), \phi(l)) : o(k, K) > o(l, L)\}$$

or

$$\{(\phi(k), \phi(l)) : o(k, K) < o(l, L)\}$$

is contained in G or $\mathbb{N}^4 \setminus G$ and $\phi(K)$ and $\phi(L)$ are rich? If these questions have affirmative answers then it may be possible to show that $X_p \otimes X_p$ is primary.

BIBLIOGRAPHY

[A1]   D. Alspach, *Another method of construction of $\mathcal{L}_p$ spaces*, unpublished manuscript (1974).

[A2]   D. Alspach, *On the complemented subspaces of X_p*, Israel J. Math **74** (1991), 33–45.

[AC]   D. E. Alspach and N. Carothers, *Constructing unconditional finite dimensional decompositions*, Israel J. Math. **70** (1990), 236–256.

[BP]   C. Bessaga and A. Pełczyński, *Spaces of continuous functions IV*, Studia Math. **19** (1960), 53–60.

[BRS] J. Bourgain, H.P. Rosenthal and G. Schechtman, *An ordinal L^p-index for Banach spaces with an application to complemented subspaces of L^p*, Annals of Math. **114** (1981), 193–228.

[CL]   P. Casazza and B. Lin, *Projections on Banach spaces with symmetric bases*, Studia Math. **52** (1974), 189–193.

[F]   G. Force, *Constructions of $\mathcal{L}_p$-spaces, $1 < p \neq 2 < \infty$*, Dissertation, Oklahoma State University, Stillwater, Oklahoma, 1995, Available from the Banach space archive (ftp.math.okstate.edu:/pub/banach/) as forcescriptLp.tex.

[JO]   W. B. Johnson and E. Odell, *Subspaces and quotients of $l_p \oplus l_2$ and X_p*, Acta Math. **147** (1981), 117–147.

[JSZ] W. B. Johnson, G. Schechtman, and J. Zinn, *Best constants in moment inequalities for linear combinations of independent and exchangeable random variables*, Ann. Prob. **13** (1985), 234–253.

[LP]   J. Lindenstrauss and A. Pełczyński, *Absolutely summing operators in $\mathcal{L}_p$-spaces and their applications*, Studia Math. **29** (1968), 275–326.

[LR]   J. Lindenstrauss and H. P. Rosenthal, *The $\mathcal{L}_p$ spaces*, Israel J. Math. **7** (1969), 325–349.

[LT]   J. Lindenstrauss and L. Tzafriri, *Classical Banach spaces*, Lecture Notes in Mathematics 338, Springer-Verlag, Berlin, 1973.

[LTI] J. Lindenstrauss and L. Tzafriri, *Classical Banach spaces I, Sequence spaces*, Springer-Verlag, Berlin, 1977.

[LTII] J. Lindenstrauss and L. Tzafriri, *Classical Banach spaces II, Function spaces*, Springer-Verlag, Berlin, 1979.

[R]   H. P. Rosenthal, *On the subspaces of L_p ($p > 2$) spanned by sequences of independent random variables*, Israel J. Math. **8** (1970), 273–303.

[R2]   H. P. Rosenthal, *On the span in L^p of sequences of independent random variables (II)*, Proceedings of the Sixth Berkeley Symposium on Mathematical Statistics and Probability, Vol. 2, University of California Press, Berkeley and Los Angeles, 1972, pp. 149–167.

[S]   G. Schechtman, *Examples of $\mathcal{L}_p$ spaces ($1 < p \neq 2 < \infty$)*, Israel J. Math. **22** (1975), 38–147.

[TJ]   N. Tomczak-Jaegermann, *Banach-Mazur distances and finite-dimensional operator ideals*, Pitman Monographs and Surveys in Pure and Applied Mathematics, vol. 38, Longman.

[W]   P. Wojtaszczyk, *Banach spaces for analysts*, Cambridge Studies 25, Cambridge University Press, Cambridge, 1991.

[W2] P. Wojtaszczyk, *Uniqueness of unconditional bases in quasi-Banach spaces with applications to Hardy spaces, II*, to appear, Israel J. Math.

[Woj] M. Wojtowicz, *On the Cantor-Bernstein type theorems in Riesz spaces*, Indag. Math. 50 (1988), 93–100.

[Wo1] J.Y.T. Woo, *On modular sequence spaces*, Studia Math. **48** (1973), 271–289.

[Wo2] J.Y.T. Woo, *On a class of universal modular sequence spaces*, Israel J. Math. **6** (1975), 193–215.

OKLAHOMA STATE UNIVERSITY, DEPARTMENT OF MATHEMATICS, STILLWATER, OK 74078
E-mail address: alspach@math.okstate.edu

Editorial Information

To be published in the *Memoirs*, a paper must be correct, new, nontrivial, and significant. Further, it must be well written and of interest to a substantial number of mathematicians. Piecemeal results, such as an inconclusive step toward an unproved major theorem or a minor variation on a known result, are in general not acceptable for publication. *Transactions* Editors shall solicit and encourage publication of worthy papers. Papers appearing in *Memoirs* are generally longer than those appearing in *Transactions* with which it shares an editorial committee.

As of December 31, 1998, the backlog for this journal was approximately 5 volumes. This estimate is the result of dividing the number of manuscripts for this journal in the Providence office that have not yet gone to the printer on the above date by the average number of monographs per volume over the previous twelve months, reduced by the number of issues published in four months (the time necessary for preparing an issue for the printer). (There are 6 volumes per year, each containing at least 4 numbers.)

A Copyright Transfer Agreement is required before a paper will be published in this journal. By submitting a paper to this journal, authors certify that the manuscript has not been submitted to nor is it under consideration for publication by another journal, conference proceedings, or similar publication.

Information for Authors and Editors

Memoirs are printed by photo-offset from camera copy fully prepared by the author. This means that the finished book will look exactly like the copy submitted.

The paper must contain a *descriptive title* and an *abstract* that summarizes the article in language suitable for workers in the general field (algebra, analysis, etc.). The *descriptive title* should be short, but informative; useless or vague phrases such as "some remarks about" or "concerning" should be avoided. The *abstract* should be at least one complete sentence, and at most 300 words. Included with the footnotes to the paper, there should be the 1991 *Mathematics Subject Classification* representing the primary and secondary subjects of the article. This may be followed by a list of *key words and phrases* describing the subject matter of the article and taken from it. A list of the numbers may be found in the annual index of *Mathematical Reviews*, published with the December issue starting in 1990, as well as from the electronic service e-MATH [**telnet e-MATH.ams.org** (or **telnet 130.44.1.100**). Login and password are **e-math**]. For journal abbreviations used in bibliographies, see the list of serials in the latest *Mathematical Reviews* annual index. When the manuscript is submitted, authors should supply the editor with electronic addresses if available. These will be printed after the postal address at the end of each article.

Electronically prepared papers. The AMS encourages submission of electronically prepared papers in $\mathcal{A}_\mathcal{M}\mathcal{S}$-TeX or $\mathcal{A}_\mathcal{M}\mathcal{S}$-LaTeX. The Society has prepared author packages for each AMS publication. Author packages include instructions for preparing electronic papers, the *AMS Author Handbook*, samples, and a style file that generates the particular design specifications of that publication series for both $\mathcal{A}_\mathcal{M}\mathcal{S}$-TeX and $\mathcal{A}_\mathcal{M}\mathcal{S}$-LaTeX.

Authors with FTP access may retrieve an author package from the Society's Internet node `e-MATH.ams.org` (130.44.1.100). For those without FTP access, the author package can be obtained free of charge by sending e-mail to `pub@ams.org` (Internet) or from the Publication Division, American Mathematical Society, P.O. Box 6248, Providence, RI 02940-6248. When requesting an author package, please specify $\mathcal{AMS}$-TEX or $\mathcal{AMS}$-LATEX, Macintosh or IBM (3.5) format, and the publication in which your paper will appear. Please be sure to include your complete mailing address.

Submission of electronic files. At the time of submission, the source file(s) should be sent to the Providence office (this includes any TEX source file, any graphics files, and the DVI or PostScript file).

Before sending the source file, be sure you have proofread your paper carefully. The files you send must be the EXACT files used to generate the proof copy that was accepted for publication. For all publications, authors are required to send a printed copy of their paper, which exactly matches the copy approved for publication, along with any graphics that will appear in the paper.

TEX files may be submitted by email, FTP, or on diskette. The DVI file(s) and PostScript files should be submitted only by FTP or on diskette unless they are encoded properly to submit through e-mail. (DVI files are binary and PostScript files tend to be very large.)

Files sent by electronic mail should be addressed to the Internet address `pub-submit@ams.org`. The subject line of the message should include the publication code to identify it as a Memoir. TEX source files, DVI files, and PostScript files can be transferred over the Internet by FTP to the Internet node `e-math.ams.org` (130.44.1.100).

Electronic graphics. Figures may be submitted to the AMS in an electronic format. The AMS recommends that graphics created electronically be saved in Encapsulated PostScript (EPS) format. This includes graphics originated via a graphics application as well as scanned photographs or other computer-generated images.

If the graphics package used does not support EPS output, the graphics file should be saved in one of the standard graphics formats—such as TIFF, PICT, GIF, etc.—rather than in an application-dependent format. Graphics files submitted in an application-dependent format are not likely to be used. No matter what method was used to produce the graphic, it is necessary to provide a paper copy to the AMS.

Authors using graphics packages for the creation of electronic art should also avoid the use of any lines thinner than 0.5 points in width. Many graphics packages allow the user to specify a "hairline" for a very thin line. Hairlines often look acceptable when proofed on a typical laser printer. However, when produced on a high-resolution laser imagesetter, hairlines become nearly invisible and will be lost entirely in the final printing process.

Screens should be set to values between 15% and 85%. Screens which fall outside of this range are too light or too dark to print correctly.

Any inquiries concerning a paper that has been accepted for publication should be sent directly to the Editorial Department, American Mathematical Society, P. O. Box 6248, Providence, RI 02940-6248.